Living in a man-made world

Gender assumptions in modern
housing design

Marion Roberts

London and New York

First published in 1991
by Routledge
11 New Fetter Lane, London EC4P 4EE

Simultaneously published in the USA and Canada
by Routledge
a division of Routledge, Chapman and Hall, Inc.
29 West 35th Street, New York, NY 10001

© 1991 Marion Roberts

Typeset by Michael Mepham, Frome, Somerset, BA11 2HH
Printed and bound in Great Britain by
Biddles Ltd, Guildford and King's Lynn

British Library Cataloguing in Publication Data
Roberts, Marion, 1951 –
    Living in a man-made world: gender assumptions in modern
    housing design.
    1. Great Britain. Houses. Architectural design – Sociological
    perspectives
    I. Title
    307.336

Library of Congress Cataloging in Publication Data
Roberts, Marion.
    Living in a man-made world: gender assumptions in modern
    housing design / Marion Roberts.
        p.   cm.
    Includes bibliographical references.
    1. Architecture and women – Great Britain.   2. Architecture.
    Domestic – Great Britain.   I. Title.
    NA2543.W65R64   1990
    728'.0942–dc20   90-35824
                                                                        CIP

ISBN 0-415-05747-7
ISBN 0-415-03237-7 (pbk)

Contents

Figures and tables

Preface

This book is the culmination of a long project which started nearly ten years ago. At that time I was working as an architect, strongly committed to what was then a rather unfashionable form of practice, known as community architecture. As I was working on housing many of the user-clients of the schemes I was involved in were women. The lack of women in the profession was such that it had been a huge culture shock to me when I had started at architectural school. The transition from all-girls grammar school to almost all-male university department had not passed without trauma. The mis-match between practitioners and recipients was all too obvious.

I was not alone in these feelings of confusion. Other women in the profession had also been touched by, or active in, the Women's Movement and wanted to extend the ideas and practice of feminism to architecture. A successful conference on the theme of 'Women and Space' was organised in 1979: following this a group of us got together to write a book on the topic of women and architecture. *Making Space* was finally published by Pluto Press in 1984 under the collective authorship of the Matrix Book Group. Matrix also became an architectural practice in which some of the book group members, but not myself, were involved. Matrix has since flourished and after some ups and downs is now in a healthy state. Another all-women feminist research and resource centre was also formed – The Women's Design Service – and is still providing an outstanding contribution. They produce a regular journal, *WEB*, which deals with all aspects of women and the built environment.

The take-over of many local authorities by socialist–feminist administrations provided a boost to the consideration of women's issues in a practical sense. Some women's committees took up housing and planning and started to offer advice and guidance on good practice. A fruitful association between the Greater London Council and the Polytechnic of Central London School of Urban Planning resulted in some useful research on the subject. However, the sustained attack which has been mounted on local authorities by central government and some sections of the Press has resulted in a dissolution of many of these initiatives. Intellectual interest has not declined, however, and many interesting articles and papers are still being produced.

This brief summary supplies the context to my work. While this was happening I had begun a PhD: after some searching for a research grant I got started and finally completed in 1986. This book is based on my thesis work and is intended to be of use to students of housing, planning, architecture and women's studies: and of interest to architects, planners and politicians. It is the first full-length book on women and design based on English data. As such it is an exploration and bears the unevenness and crudity of a first attempt. I hope that it will provide a marker: a point from which others will take off to map out and explain this fascinating and important topic.

Acknowledgements

I have many diverse people and institutions to thank for their help and encouragement in carrying out the research on which this book is based and in the writing of the book itself. First and foremost I should like to thank the women of the Matrix book group whose early discussions provided the inspiration for the entire project. Madeleine Arnot and the Open University set me on the road to academic research and without them I would not have started.

Hilary Land and Alan Lipman supervised my PhD research with skill and tact over personal as well as intellectual difficulties. The Science and Engineering Research Council through the Welsh School of Architecture, UWIST, provided a much-needed research studentship. Sheila Allen, as external examiner, gave a detailed and thorough set of criticisms. I regret that I have been unable to answer them in full and hence to meet her high standards. Nevertheless they have improved the book greatly.

Without a research award from the Board of Architectural Education of the Architects' Registration Council of the United Kingdom, I should never have had time to write the book. I would also like to thank CPH Architects for their flexibility in employing me on a four-day week whilst I was writing it. Ian Cooper and Jane Darke provided invaluable comments on the first draft, which gave both stimulus and criticism. I should also like to thank the audiences at various talks which I have given on aspects of the book for their comments and advice. Kenneth Player helped at the last minute by typing the bibliography. On a more personal level I would like to thank Tricia French for her encouragement and Piers Corbyn for his support and practical help.

I should like to thank HMSO for their permission to reproduce three diagrams (Alternative Ways of Living) from *The Dudley Report (1944)* and the Central Office of Information for their permission to reproduce Table A1 which was compiled from *The British Household (1949)*. Baroness Denington and Mr B. L. (Beak) Adams have also kindly given permission to publish excerpts from my interviews with them. The staff at the Greater London Records Office and the Public Records Office offered assistance in locating unpublished sources. My heartfelt thanks go to the tenants on the Crescent Estate who gave up their time for my interviews. Finally, I must make the customary assurance that despite everything, the errors and omissions in the text are mine alone.

Chapter one

Introduction

> Among the working class the *wife* makes the home.... The working
> man's wife is also his housekeeper, cook and several other single
> domestics rolled into one; and on her being a managing or
> mismanaging woman depends whether a dwelling will be a home
> proper, or a house which is not a home.
>
> (Thomas Wright, engineer, 1868 cited in Cockburn 1983: 34)

The above quotation illustrates attitudes which, although they might not
be stated explicitly, are commonplace now, over a century later. So commonplace are these ideas that they have pervaded not only the social
relations of home but the very building of houses themselves.

This book is written from a feminist perspective and is about housing
design. Divisions of gender have had a profound effect on British economy and culture. Recent feminist studies have articulated these divisions
in a wide number of fields – in education, in the arts, in the legal system,
in waged work and in the domestic sphere. Little attempt has been made
to explore housing itself though, the bricks and mortar of home, a key site
of gender division and subordination.

There are two reasons for this apparent neglect. The first is the overwhelming domination by men of the building-design professions and the
construction industry. Approximately 7% of architects are women (Walker 1989). Because the numbers of women are so small, stereotyped
responses are common. 'Ah, a woman architect – now here's someone
who knows how to design a kitchen' runs a conventional vein of comment. For this reason some women architects purposefully steer
themselves away from housing design and its associations with domesticity and keep to public and commercial work, in an attempt to prove
their equality.

With these kinds of pressures it is tempting to treat the topic in a super-
ficial way. For example, the argument might be constructed that if only
there were more women in the design professions and in the building
industry then any problems which might be experienced by women would

be solved. Yet more women professionals would not affect the price of housing, its location, nor indeed the cultural importance which is placed on having a clean and tidy home. Yet each of these issues affect women's choices of work, love and independence in a deep way. The cost of buying or renting housing affects whether women can afford to set up households independently from men, either as single women, with children and, or with other women. The location of housing in relation to empoyment affects married women's choices of career development. The cultural value placed on a comfortable, clean, tidy home and the struggle to achieve and maintain this, can place strains on intimate relationships – most commonly with husbands but also with other cohabitees.

Because housing is so clearly part of society – as part of the economic system, as part of the material world and as part of culture, a detailed analysis is necessary. This has to be done, however, within a framework of ideas which makes coherent such a vast topic. There have been many histories of housing (e.g. Merrett 1979, Merrett & Gray 1982; Burnett 1980) and there is a continuous stream of discussion of housing policy in books, journals and the popular press. Few books have taken an explicitly feminist approach and these have divided equally between issues in housing policy (e.g. Watson 1986; Brion & Tinker 1980) and those in design (e.g. Hayden 1981; Matrix 1984; Attfield & Kirkham (eds)1989). With such a scattered and recent literature I have had to start afresh on the topic. In order to provide a structure to the book I have drawn on four themes which address different aspects of the relationship between housing and gender.

There is also the issue of feminism to be addressed. The Women's Movement is not unanimous, if indeed it ever was. There has been no national conference since 1978, nor does there seem likely to be one. There are many views of feminism and some are at odds with each other. For example, socialist feminists have criticised radical feminists for having a fundamentalist view of biological difference (Segal 1987) and black feminists have criticised white feminists for being middle class and ethnocentric (Hooks 1984). Before considering the four themes I have used to tie this study together, I shall first discuss the feminist viewpoint that they are approached from.

Feminism – old and new

The issue which divides the Women's Movement is who or what is the enemy. This is a matter of theory, strategy and practice. In the late 1970s a split occurred between socialist feminists and radical feminists. Socialist feminists saw the dominance of men over women and class politics as acting together in the oppression of women, whereas radical feminists argued that male dominance formed the motor force behind women's subordination from which all men, of whatever class, benefited. Indeed Delphy, for example, argued that women constituted a class in themselves (Delphy 1970).

Since then there have been other splits and variations. In particular there has been an emergence of identity politics in which different groups of women have articulated and campaigned around their particular struggles. Thus women from ethnic minorities have criticised white feminists for promoting their own particular concerns – abortion rights and independence as the issues for all women. They have other issues to confront as priorities, most notably the problems of fundamentalist religions which regard women's subordination as God's will.

Lesbian women also face particular forms of difficulty because of prejudice about their sexuality. Some see a society which values heterosexual relations to the exclusion of all others as being intensely problematic and have campaigned and worked not only to forge a lesbian identity and culture but also to promote positive images for homosexual relations, both male and female. Any advances which they might have made have been set back by the backlash against them, in that Clause 28 of the Local Government Act 1988 was enshrined in statute in order to prevent the 'promotion' of homosexual relations, that is the assertion of homosexual relations as having equal value as heterosexual relations.

Not all lesbians have identical political positions, though. Some would see themselves as placed more firmly within a socialist tradition which places an emphasis on class. They would reject some separatist arguments and assertions. Thus although many campaigns have been fought on the basis of identity politics, not all members of those groups necessarily agree with all those campaigns.

In the late 1970s and 1980s the Women's Movement has become more diverse and more particular. Campaigns have been mounted against nuclear weapons, against pornography and against violence against women. Solidarity networks have been formed with women's organisations in Nicaragua, Palestine, India and other developing countries. Although organisations such as Women For Socialism attempt to draw together these diverse strands into unity there are theoretical differences on the aforementioned issues.

The extent to which men and women are regarded as being innately different forms one basis for disagreement. The extent to which men have an innate propensity to rape, to make war and to subjugate women is cause for debate. One separatist position is that male power is exercised through fear, and pornography and rape are extreme manifestations of men's control over women. This male propensity can only be challenged by women living and organising separately.

Another position to which I subscribe is that rape and pornography are extreme manifestations of imbalances and inequalities which exist in society as a whole. These inequalities are enacted economically through the wage system, the ownership of property, the sexual division of domestic labour (i.e. women having major responsibility for housework and child rearing) and are legitimated ideologically by a variety of means, for example, through advertising and the educational system. State intervention acts on both a material and ideological level by disadvantaging women materially and providing reasons and justifications for treating women as subordinate. Obviously there are different views of the state, which have been summarised more fully by others elsewhere (George & Wilding 1978). My own view is that the state in Britain is controlled largely by those who have access to power and wealth and yet is still susceptible to parliamentary and extra-parliamentary action. The state therefore has a dual nature and almost each benefit that it administers has an element of control over the working class and gain for the working class. Men benefit materially from inequalities of gender, through enjoying access to greater resources and having less responsibility, but suffer emotionally and spiritually from a loss of real relationship with equals and loss of contact with children.

This book, then, is written from what might be described as 'an old-fashioned' feminist viewpoint. My concerns are with male stereotyped views of women, with inequalities in state provision, with inadequate provision for childcare. There is also an idealist strand, a belief that if the mechanisms for inequality are exposed, then these can be changed and from that, men and women will change to make a better society.

Whilst it is easy to discuss 'women' as opposed to a separate category of 'men' in order to achieve conceptual clarity, this is an over-simplification of most women's lived experience. There are inequalities between women of race and class, and a variety of experience in terms of sexual orientation, long-term relationships and motherhood. It is therefore inaccurate to portray the interests of a white single woman in a professional job as being coincident with the grandmother in an extended family from an ethnic minority. In this book I have attempted to differentiate between groups of women and to analyse differences of class-status. Differing theoretical approaches to class will be referred to in more detail further on in this introduction.

Although I have talked about women in this introduction, gender forms a central concept in this book. By this, I mean the belief that femininity and masculinity are constructed and imposed by society and that there is not, nor should there be any hierarchy between women and men. Biology is given, gender is learned.

Feminists such as McIntosh and Barrett have argued that marriage is an oppressive institution (Barrett & McIntosh 1982) which cuts women off from wider contacts and reinforces women's subordinate position in society. Although some of their criticisms of contemporary family life will be taken up in this book, I disagree with their objections to family life as such. It would seem to me that it has been a failure of the contemporary Women's Movement not to adequately describe or theorise the dual nature of familial relations. On the one hand, family life can be a source of severe distress and danger – through domestic violence and poverty. More generally, women's secondary position in the labour market and dependence on men in welfare benefits and other legal matters is reinforced by expectations of family life. On the other hand, marriage and child rearing can provide an immense source of non-material happiness – a valuable counterpart to the wage-earning nexus of a capitalist society.

Because of this dual nature of marriage and family life, I would argue for reforming their arrangements rather than revolutionising them. In common with Barrett and McIntosh I would suggest that there should be greater freedom to form of all sorts of different household types. Women, and particularly working-class women who are disadvantaged financially, should not be forced by circumstance into unwelcome relationships. They should be free to enter or leave cohabitation with men and women as they wish. Economic necessity should not provide the motor force of domestic arrangements. Furthermore, equality within marriage and relationships similar to marriage is also vital and should be struggled for. Since the family is an intrinsic part of a society in which male, paternalistic authority has dominance (Gittins 1985) this means that other social relations and practices also have to change.

This book is also based on white women's experiences and expressed needs. This is partly because the book is based historically and immigration by black people to Britain in significant numbers has been relatively recent. Some racist practices by a local authority housing department were observed and have been noted, although the racist comments of some white tenants have been excluded. The reason for this was so as not to give credibility to these opinions by printing them. However, that they were aired at all illustrates the extent of divisions between women. The housing needs of women from ethnic minority groups do require research and action.

Similarly, there is no reference to the needs of other groups – disabled people and lesbians. Throughout the book I have emphasised the need for women to gain access to housing independently of men and to get employment which fulfils their potential. If this were achieved, it would go a long way towards meeting the housing needs of lesbians, although this would not be a complete solution to all problems.

These, then, are the concepts and ideas which have been used in writing this book. In order to give coherence to my work and because there was no ready-made theoretical framework, I have organised my discussion around four themes. These will be explained next.

Social reproduction and housing form

The concept of reproduction derives from Marxism and is used to describe the processes of renewal within society outside the production of commodities. It has been used by Williams to discuss the way in which culture supports the domination of an elite but at the same time retains a vitality of its own (Williams 1973).

The concept of reproduction has three identifiable components. The first is that through having a home, workers are fed, clothed and generally prepared for work. Motivation to gain employment is also given meaning through the home and the personal relationships nurtured in the home. Reproduction also has a biological, material sense, for workers are literally reproduced through child birth and reared in the home. Finally children are socialised into taking their places in society and in this sense ideology is reproduced. Of course parents can and do inculcate their children with subversive ideas as well as ideas which support the ruling class. Dominant ideas about 'home' and 'family' contribute to these processes of ideological and physical renewal.

There has been debate amongst Marxist thinkers about the precise relationship between productive and reproductive forces. Williams argues that reproductive forces are part of the base of society, helping to forge a dominant culture which strengthens and reinforces capitalism. The family plays a key part in reproduction, for it is within the family that children are socialised and learn how to become workers.

Whilst some authors have suggested that the current form of nuclear family is therefore the base unit of capitalism and in order to overthrow capitalist relations the nuclear family must be revolutionised (Zaretsky 1976), that is not the inference which I wish to draw here. Rather, I would suggest that in the total process of society inequalities within the nuclear family help to support aspects of capitalist relations and the dominance of masculinity.

In a capitalist system the rules of the market dominate other considerations of provision. Thus in the nineteenth century, rented housing was built by the aristocracy, small entrepreneurs and other businessmen and let to whoever could afford to rent it. Middle-class philanthropists intervened to provide housing for the working classes both as an altruistic gesture and for a modest return on investment (Tarn 1973). In the early twentieth century the state set up a national programme of building houses to rent, but again with a duality of purpose: to provide for need and to stave off the threat of revolution (Orbach 1977).

Later in the twentieth century, the state supported owner occupation as well as council housing. Again as well as helping to meet need there was an ideological edge, certainly from the Conservative Party's viewpoint since they believed that owner occupation would prevent a drift to socialism (Merrett & Gray 1982). Thus housing is moulded by the processes of production and reproduction in society. Privately rented housing, council housing and owner occupied housing all form part of the productive process in society, because although council housing is provided by the state, it is still built by privately owned builders and funded by loans that derive from the City. Lack of housing also means that an individual is marginalised from the productive processes, for without a house it is difficult to find a job.

Housing forms part of the reproductive processes in society in that the provision of both owner-occupied and council housing have bolstered the male-dominated nuclear family. This last assertion has been argued by Watson (1986) and further evidence will be provided in this book. Furthermore, since the focus of this book is gender and women, it will be argued that the state has intervened at particular times and in particular ways in order to support different roles for married, working-class women – as waged workers, as home makers or as mothers. These interventions have affected the design of of housing in terms of its physical shape and size.

This book concentrates on working-class housing because apart from housing built for the very rich, that is the sector over which architects have had most influence. As an architect myself, it is this field of practice which I find most interesting. Moreover, since local authority housing is provided by public bodies, it is easier to gain access to documentary records. Finally, it is the sector in which there is most need and in which most remains to be done. The first two chapters of this book will concentrate on the theme outlined above.

Housing and gender

The location and cost of housing is as important to considerations of gender as its style, shape and size. Because women occupy a secondary position in the labour market to men, their earning power is limited. Women's employment possibilities are also inhibited by their domestic responsibilities which in turn also decreases their wage-earning capacity.

The reasons why women occupy a secondary position in the waged labour market have been much debated. Beechey has argued that women constitute a reserve army of labour, in the Marxist sense (Breugel 1986). By this she means that women are drawn into the waged labour market in times of economic boom and expelled from it into unpaid work at home in times of slump. Breugel has qualified this assertion, reporting that in the recession of the early 1980s women were not made redundant from the waged labour force. Whilst women's full-time employment dropped during this period in line with men's, women's part-time unskilled employment actually rose.[1]

The sexual segregation of the workforce is readily observable, with many industries which are either male dominated to almost the exclusion of women, such as, for example, the building trades, and others which are almost entirely female dominated, such as secretarial work. Phillips and Taylor have shown how skill is the subject of negotiation. In certain instances male definitions of what constitutes skilled work have predominated, to the detriment of certain industrial processes carried out by women (Phillips & Taylor 1980). Cockburn has taken this argument deeper and has explained in her study of the printing trade, how men were able to define physical aspects of the process of compositing and give them a mystique which worked against the employer's interests of profitability and women's interest in equality (Cockburn 1983).

One of the main explanations for occupational segregation is the ideology of domesticity and women's responsibilities for housework and childcare. Beechey has criticised a number of studies and indeed her own previous arguments for drawing too uncritically on this explanation. She makes the point that there is a significant difference between ideological assumptions about women's responsibilities for housework and childcare and the realities of those responsibilities (Beechey 1986). She suggests that in some industries, such as garment making, occupational segregation has come about because employers have actively used changes in technology to de-skill and casualise female labour. In other types of industry, such as local government, she suggests that women tend to occupy certain posts as an extension of their familial roles.

The concept of the family wage is also used by employers and trade unionists to marginalise women workers, regardless of their circumstances. Land has shown how this concept has been used to suggest that women have no right to a job, the argument being that a man's wage should be sufficient to provide for his family of dependants and that women by being employed are 'taking away' jobs from men (Land 1980).

The causes of women's secondary place in the waged labour market are therefore only partially known. Planning policies play their part in reproducing this division of gender. Lewis has shown how a new town made specific efforts to attract employers who could provide jobs for women – most often low-paid and unskilled (Women and Geography Study Group 1984). In this way there was a continuous interchange between the pressures of the market encouraging employers to keep wages low, and family ideology encouraging women to put domestic responsibilities first.

Since the Second World War the state has intervened substantially in the relative locations of housing and industry. In Chapter four I shall examine the assumptions which planners and architects made about working women and men and the nuclear family in relation to planning.

Hall has described how in the mid-nineteenth century female employment in the Lancashire cotton towns raised fears for the family and the spectre of a reversal of sex roles. These fears were expressed by the socialist, Engels, who worried that the employment of women and children at low wages would drive men out of work. Such a situation, he suggested, would either dissolve the family or, in many cases, the family would be 'turned upside down'. Men would have to stay at home and look after the children and women would be out at work. This, he thought, would unsex both sexes, depriving men of their masculinity and women of their femininity (Hall 1982). Chapter four will examine how this fear of role reversal and undermining of masculine superiority was not confined to the nineteenth century: it also formed a strong undercurrent to post-war plans for decentralising the capital.

Although state intervention in the provision of housing meant that better standards of housing could be provided than they would normally be able to afford under market conditions, because resources were and are limited, assumptions were made about the type of household which was regarded as being most in need. Chapter three will explore how 'nuclear' families were prioritised over all other household types,

State intervention in raising housing standards meant that as in other fields of social welfare, the state was making an intervention into the class structure. Although the relative importance of home has changed historically in terms of definitions of personal wealth – for example, Athenians in the second and third century BC lived in cramped dark houses (Walker 1983) – nevertheless housing does play an important part in determining social status. In the next section I shall look at the third of my themes, which is housing and class.

Housing and class

In the last decade considerations of tenure have dominated discussions of housing and class. Since 1979 a dominant theme of housing legislation has been to encourage owner occupation to the maximum extent. Council tenants have been persuaded to buy their council houses and flats with discounts of up to 60% and 70% of the price of their property. The encouragement of owner occupation did not begin in 1979, however. It has been a feature of both Conservative and Labour administrations since the 1920s: the difference being that since 1979 the balance has been tilted towards owner occupation to the detriment of other tenures.

Financial institutions such as building societies have been encouraged to diversify and attract more customers with more flexible mortgage arrangements. Other institutions, such as banks, have also entered the market for home loans. Meanwhile the rate of new council house and housing association building has declined. Subsidies have been withdrawn from council housing and housing associations and tenure legislation changed with the balance going more in favour of the landlord. The result of this effort has been that now 64% of dwellings are in owner occupation in England and Wales (Central Statistical Office 1989). This has shifted accepted meanings and values of owner occupation in terms of perceptions of class and status.

Since owner occupation has become the majority tenure academic debate has proliferated over its precise relation to class structure.[2] Whether owner occupation leads to a diminution in working class militancy, whether owner occupiers form a class in themselves and if owner occupation supports a greater emphasis on privacy and home centredness have been extensively debated. These debates have concentrated on the issue of tenure to the virtual exclusion of design.

From a historical perspective it is important to remember that large-scale owner-occupation is a relatively recent phenomenon. At the end of the nineteenth century over 90% of dwellings were privately rented. In 1938 58% of dwellings were privately rented; 10% were owned by local authorities and 32% were in owner occupation. By 1953, 54% of dwellings were privately rented, 18% were rented by local authorities or New Town corporations and 28% were owner occupied. In 1960 the proportion of owner-occupied dwellings had moved up to 44% and 25% were rented by councils or other public bodies. In 1975 owner occupation reached a majority of 55%, with public renting at 29% and private renting at 16% (Merrett 1979; Murie 1983).

Thus for most of the century, status divisions have not only been drawn on tenure lines, but also on other aspects of housing. For the Victorian middle classes, size, location and privacy were important factors in definitions of status. By privacy there is not only the privacy of the individual house from the road or street, but the privacy of the household within the house.

That the importance placed on these boundaries varies between classes and period is delightfully illustrated in a collection of papers edited by Lewis, where Dyhouse's account of the rituals of an Edwardian upper-middle-class household provides a strong contrast to Gittins' reconstruction of the life of working women in Devon. In the upper-middle-class home servants were separated from family and the lavatory could not be mentioned by name. Illegitimate birth was unthinkable (Dyhouse 1986). In the rural town women lived in households where the personnel changed gradually over the years from mothers and half-brothers and sisters to husbands and children to lovers and illegitimate children and illegitimate grandchildren and half-brothers and sisters (Gittins 1986).

Status rests not only on the physical appearance of the house and the social relations of those within it but also on the manner of housekeeping. Ginsburg cites local authority manuals written in the 1950s which exhorted tenants to keep their homes clean and tidy (Ginsburg 1979). That cleanliness and tidiness are not absolute values but are ways of ordering the world and are therefore relative, has been illustrated by Douglas. She shows how objects which are not dirty in themselves may become dirty through their location, as, for example, shoes on a table (Douglas 1966). This notion can also operate in reverse, so that symbols of cleanliness may make a place 'clean' or 'respectable'. White doorsteps and spotless lace curtains were classic markers of status in working-class by-law terraced houses, providing symbols of the moral rectitude of their residents.

Housekeeping, then, provides a point where the intersection of gender and class-status may be discerned. This will be explored in Chapter seven of this book. Gender has tended to be ignored in discussions of housing class, despite policy makers' references to a householder's natural desires 'to have independent control of the home that shelters him and his family' (DoE 1971). Rose has pointed out that wives of householders living in owner-occupied suburbia may be isolated (MacKenzie & Rose 1983). Apart from Williams' (1987) survey of housing, class and gender 1700–1900, there has been little examination of this aspect of the topic. Yet in any study of gender and housing design neither gender nor class can be ignored.

Gender and class

The connections between gender divisions and class divisions have been the subject of debate for the last decade. As yet no conclusive theory has evolved to connect the two modes of analysis: nor has there been an interlinked set of concepts.

There are difficulties in attempting to place women in the class structure. How, for example, can women who work in white-collar jobs married to semi-skilled workers be considered? What about the cleaner married to a clerk – is she working class or middle class? To place women in the same class as their husbands ignores these differences and problems.

The consideration of women as workers has also led to difficulties, as was noted in the previous section. Because women tend to be segregated into particular jobs and industries and overall generally earn less than men comparison with, say, the Registrar General's classification of class-status groups is difficult. Furthermore, a significant proportion of the work which women do is unpaid: as mothers, housewives, carers for the sick and elderly.

11

Hartmann (1979) argued that it was impossible to combine Marxist class analysis (which in itself differs from sociological analyses of class-status divisions) and feminist accounts of women's subordination. She suggested that previous attempts at doing this had tended to subordinate the importance of gender divisions to class.

Delphy took the argument further and proposed, using an analysis of French peasant families, that women in themselves constituted a class. Her argument rested on the domestic economy of provincial French agricultural families where she suggested that women's labour was appropriated by men. Although her arguments were challenging it was difficult to apply them to British society in which agriculture has been industrialised for over two centuries. Delphy's analysis is also difficult to apply to the situation of the wives of middle-class men who employ other women to do their domestic labour (Delphy 1970).

'Traditional' Marxist arguments, first proposed by Engels, have assumed that the family is a residual organisation, based on a feudal mode of production and that industrialisation drew women out of the family into the labour force. Women's liberation then rests on proletarianisation, that is, women being drawn into the waged labour force and women's unpaid work being taken over by socialised forms of domestic labour such as nurseries and restaurants. A socialist revolution would emancipate all workers which would include women.

Recent historical studies have since suggested that this is a false line of argument. Pahl has summarised recent historical evidence asserting that it is only in relatively recent times that women have withdrawn from the labour force. He suggests that from the end of the eighteenth century onwards it became a sign of status for non-aristocratic women not to engage in wage-earning work outside the home. This withdrawal affected wives of working-class men for sixty years from the mid-nineteenth century to the early decades of the twentieth century. Pahl asserts that women have evolved a series of domestic strategies at different periods and in different classes for coping with maintaining a family financially and doing domestic work (Pahl 1984).

These historical studies have shown how interconnected the two spheres, the public world of work and the private world of home, were. Even in medieval times there was a sexual division of labour in the agricultural work which men and women peasants undertook in the fields and that which they did in the household. As Tilly and Scott have shown through their survey of women's work in the nineteenth and twentieth centuries in Britain and France women's working lives have been shaped and formed through their familial responsibilities. Married women from what they termed the popular classes often moved from full-time work to a combination of homeworking and housework and childcare, to part-time work and back to full-time work in the course of a life time (Tilly & Scott 1978).

Furthermore, family historians have dispelled the myth that the 'nuclear' family, that is a household consisting of parents and dependent children is either a remnant of feudal society or a product of industrialisation. Anderson concludes, in an overview of his own and others' work, that over most of Europe, from the fourteenth century onwards the most common form of household was that of a nuclear family. Additional household members tended to be servants rather than kin (Anderson 1979). Thus as was discussed in the first section of this chapter, the notion that the 'family' in itself is the cause of women's subordination from a previous golden era appears to be untenable.

What does emerge from historical snapshots of different classes and eras is that the nature of women's work, paid and unpaid has varied considerably over time and in terms of the material circumstances of the household. The degree of status attached to employment has also varied. Thus Roberts, in her accounts of Tyneside in the period before 1940 reported married women pitying widows and deserted wives who had to go out to work, feeling that their lives were harder as a result (cited in Pahl 1984). This is because the burden of domestic work was much greater then in terms of sheer physical effort and this when combined with paid employment became very hard indeed.

These variations and shifts have led social historians such as Lewis and Davidoff and Hall to argue that gender divisions and class-status divisions are mutually interdependent and are created at the same time (Davidoff & Hall 1987; Lewis 1985). This is more of an observation than a theory and suggests that an understanding of the relationship between gender and class might proceed through a series of vignettes of different epochs and classes. This is the approach taken in this book and although five chapters focus on the period following the Second World War, two chapters form historical surveys.

Housing provides a fascinating intersection of both gender and class-status systems. It is unfortunate that social scientists have tended to adopt a cleavage between public and private in their own work, studying either employment or domestic life rather than the interaction between the two. Some recent work has tried to address this problem and to avoid these dichotomies.

Class is a slippery concept to define. Whilst the Registrar General's divisions of the population into status groups according to occupation is useful, it does not always accord with popular definitions of divisions within the working class and middle class. Marxist definitions of class also differ. Throughout this book the divisions referred to will be of class-status. The concept will be thoroughly explored in Chapter seven where tenants' own perceptions of themselves and of others are considered.

Housing has also been subject to state intervention and therefore prey to ideologies of both class and gender as are other aspects of the welfare system. Land has demonstrated the significance of familial ideology which casts women in a dependent role to the taxation and social security systems (Land 1979). Throughout this book the assumptions which policy makers, architects and planners made about the wives and daughters of working-class men will be examined.

In conclusion

This book is breaking new ground in examining the assumptions and ideas of the policy makers and designers of housing and the perceptions of some of the 'consumers'. As such it is an exploration rather than a theorem: an outline proposal rather than a working drawing.

Because the potential field is so large it would be difficult to survey, say, both owner-occupied housing and council housing in the post-war decades and to come up with any meaningful conclusions without a decade of work. Because of the constraints of time and money this book is based on a study which concentrated on one particular period, the immediate post-war decade and on state housing. This period and tenure was chosen because I thought that it would illuminate the interrelationship of policies towards state housing, gender divisions and housing. In order to focus the study further, a case study was made of the then London County Council (LCC) which, at that time, was the most advanced housing authority in Britain, if not in Europe.

Housing design does not take place in a vacuum but has a long history and tradition. I therefore thought it important both to provide a survey of pre-war working-class housing and a post-script which brought the story up to date. There are many levels of decision making in housing design, from the initial decision about location and dwelling form and density to more detailed questions of internal layout, fixtures and fittings. I have included discussion of policy at these different levels and because of my own interest as an architect, I have highlighted the views of LCC architects of this period. It would have been an anathema to have introduced a discussion of gender divisions without recording the views of women themselves. Consequently, the penultimate chapter analyses the response of first tenants to moving to a new housing estate in the 1950s.

The layout of the book is as follows. Chapter two looks at working-class housing history from the mid-nineteenth to the early twentieth centuries. Chapter three takes the history to after the Second World War and considers controversies over dwelling form. Chapters four and five examine policy makers' assumptions over design. Chapter six looks at the ideas of architects in the LCC during that period and Chapter seven gives an account of some first tenants' view of their estate. Chapter eight considers changes which have occurred until the present day and draws some final conclusions.

Chapter two

Women as homemakers I

Many of the trends and themes of twentieth-century working-class housing may be traced back to the nineteenth century. Although the social and economic position of women in the poorest sections of society changed considerably, as will be seen, the physical shape of housing developed in a manner which may be explained by reference to policies towards family life, gender relations and motherhood. These policies were not specifically directed towards these issues: rather assumptions were made about gender relations in housing, planning and public health policies.

State policies, that is, local and central government policies were neither consistent nor entirely logical in their evolution. However, throughout the latter decades of the nineteenth century and the first half of the twentieth it is possible to identify the following strands in policy. These are the support of the nuclear family as the base unit of society; the assurance of men in having greater, if not exclusive access to skilled jobs; a tentative support of motherhood and finally, but not least, an underpinning by the state of middle-class attitudes towards respectability and propriety.

Architects and planners, who, as has been previously noted, were almost all male, had considerable influence on the quality of state policies towards housing. In particular, the influence of the Garden City Movement with its proponents in Howard, Unwin and Abercrombie should not be ignored. Their vision of low-density self-contained urban development profoundly affected the evolution of British working-class housing form.

Howard was a nineteenth-century reformer who reacted strongly to the slums and squalor thrown up by the industrial revolution. It is in the physical manifestations of industrial development where I shall start my discussion of housing form and its relation to production and reproduction.

Industrialisation

Britain experienced industrialisation of its manufacturing industries earlier than other countries. This plus the colonisation of other countries helped to fuel an expansion of trade and industry. The production of goods moved outside the household or workshop into factories and processing plants. As cottage industries were undercut by manufactured goods, craftsmen and craftsmen's families were no longer able to support themselves in autonomous units, but had to seek waged employment instead.

Burgeoning industry drew in labour, from whatever source was available. Since the agricultural enclosures of the seventeenth century, country labourers had been driven off the land and forced to find work in the towns. Successive famines in Ireland also increased immigration of rural people to manufacturing centres to find work (Gauldie 1974).

From 1814 onwards the population in British towns and cities expanded at an exponential rate. From 1801 to 1911 the population of Britain grew from 9 million to 36 million. New houses were thrown up by speculative landlords. Six million houses were built between 1801 and 1911 (Muthesius 1982). Even this was not enough to contain the swelling population. Overcrowding was rife in cities and industrial towns in the middle decades of the nineteenth century. In some areas houses which had formerly been inhabited by one or two households were given over to multiple occupation. Whilst some tenants managed to keep clean, reasonable homes despite appalling overcrowding, others were dragged into miserable conditions of dirt and neglect.

In the first half of the century speculative builders constructed housing speedily and cheaply with few controls. In cities such as Liverpool terraces were tightly packed together with high densities of population (Burnett 1980). Developers were reluctant to invest in infrastructure items such as sewers and water supplies: existing services such as refuse collection could not be maintained under such stress and there were outbreaks of infectious diseases such as typhoid and cholera.

Women's employment

Industrialisation and migration from country to town caused a gradual shift in women's employment. Prior to industrial manufacture becoming the dominant method of production, the creation of goods and services took place in the family unit. Married women and young daughters took part in whatever method the family had of earning a living, whether it was a small farm or business. If there was not sufficient income available to meet the family's needs, then young daughters would be sent off to go into service, or to act as a servant-cum-apprentice in a neighbouring business. On marriage, and the majority of girls did marry or form some stable union, then wives would help husbands in their trade or with their land, or undertake some other home-based work which would provide extra income as necessary (Tilly & Scott 1978).

The replacement of craft manufacture by machine manufacture meant that work was now done outside the home rather than inside it. However, this did not mean that familial relations were destroyed: rather it meant that married women and young girls added their wages as supplementary to the main male or 'family wage'. The moving of work away from home meant that that the kind of work which married women could undertake was more limited than hitherto because of their domestic ties, it was only those casual, underpaid tasks which could be accommodated in the home. Married women's work in the nineteenth century has been ill-recorded since casual forms of work such as taking in washing and home work or 'slop' or 'sweated' work as it was known were not recorded by the Census. Alexander has provided an account of some of the numerous trades and occupations which married women undertook on a part-time or casual basis in nineteenth-century London; these trades included hawking and selling fruit and flowers, laundry work, shopkeeping, needlework, artificial-flower making and a host of other finishing trades (Alexander 1976).

Young girls still had ample employment in domestic service. Although industrialisation did open up new types of employment in factories, this was limited to particular northern towns and therefore could not be seen as a general condition (Pahl 1984). Girls working in the textile mills of the north earned unprecedentedly high wages, and their employment was such as to worry contemporary commentators that familial relations would be disturbed (Hall 1982). However, their experience was not typical of the country as a whole. Marx describes, in an extreme of a more typical situation, how a young milliner died through overwork in a stuffy, badly ventilated workshop (Marx 1887). There is also some evidence to suggest that certain factory owners, not wishing to disturb paternal authority employed whole families at once, paying the wages to the husband for distribution to his children. Miners, for example, worked in this way (Humphries 1981).

Whilst a combination of male and female wages was necessary in many manual occupations in order to keep a family above subsistence level, it was not always possible. Death rates were high in nineteenth-century cities and the death of a spouse was a not uncommon event (Gittins 1985). The plight of the poor widow provided a focus for sentimentality in popular culture.

In the absence of any effective social services apart from a punitive poor-law and workhouse system, those unable to maintain themselves or to depend upon spouses or relatives were forced to beg, become vagabonds or to take up prostitution. This latter was of particular importance for women and the phrase 'woman of the streets', meaning a prostitute, has particular resonance in terms of meaning without a house (Davidoff, L'Esperance & Newby 1976).

The cult of domesticity

Whilst working-class women toiled in stuffy workshops and in slum houses, upper-middle-class women suffered enforced leisure. Whereas in the sixteenth and seventeenth centuries the wives of merchants had been able to participate in commercial life, through their husbands, or as widows, by the early nineteenth century this situation changed.

To begin with, a new class of wealthy tradesman emerged. As Britain became a colonial power, her markets for goods, as well as sources of raw materials expanded. Machine methods of manufacture superseded craft production and created vast amounts of wealth for the owners of factories, mills and other enterprises. More sophisticated distribution networks were also required, which meant that a tier of marketing and warehousing businesses could flourish.

Hand in hand with the emergence of this new bourgeoisie came an associated ideology of gender difference. In seeking a place for themselves in the world the merchants and shopkeepers distanced themselves from what they saw as a degenerate aristocracy. The cult of evangelism also played a part in the formation of the new morality. Puritanism had been a religious tradition in Britain since the sixteenth century. Evangelism had many similarities with puritanism, with its emphasis on personal morality and conscience, hard work and gender difference.

For in evangelism the two sexes were conceived as different types of complementary being: men were thought to be fiery, worldly and active, women to be mild, gentle and inward looking. Women were thought to be the guardians of morality, their restraining influence would help to keep men on the right course. Such influence, however, was to be exerted in the home since women were not suited to the rigours of the outside world.

These notions of an acute divide between the sexes shared in the formation, some argue, of a clear physical, locational divide between the sphere of influence of each sex, taken literally between home and work. As businesses grew it no longer became possible for the family to live 'over the shop'. Specialised institutions, such as insurance firms and banks and the professions excluded women from their ranks. In order to gain prestige and comfort, members of the wealthy bourgeoisie bought or rented houses at the edge of the city (Davidoff & Hall 1987).

With the purchase or lease of these homes came a clear divide between home and work. The master of the house would leave the house for work

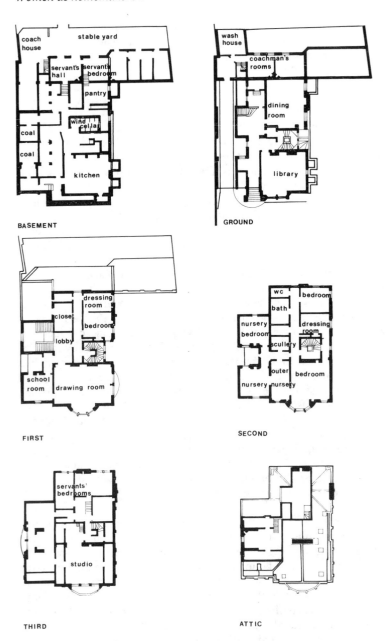

Figure 2.1 Plans of upper-middle-class house *c.* 1860

in the city, wives, daughters and other female relatives remaining within the house to run the household. Although the house was a business in one sense, in that both female and male servants were employed there, no productive work was done within it. Instead, women practised crafts which were for decoration and amusement instead of barter, such as fine needlework, embroidery, tapestry and watercolour painting.

It was not only the ideas of religious sects which supported the seclusion of middle-class women. Early Victorian popular culture produced a plethora of tracts, sermons, engravings, songs and poems which celebrated home and the role of women within it. Women were the 'angel in the house' succouring and providing a cosy nest for men to retreat to.

Homes themselves became more elaborate than in preceding centuries for the middle classes. Tables were covered with cloths, floors with carpets and every surface covered with pictures, fans and other ornaments. Meals, in contrast to the preceding century, changed from everything being served on the table at once, to have separate, distinct courses (Black 1985). Childcare also became more ordered and regulated, with nannys, nurses and governesses providing order and instruction.

House plans tended to intensify hierarchies of class, as well as gender. Figure 2.1 shows a large house designed for an upper-class family in the 1860s in London. This plan is unusual in that it contains a studio for an artist. Nevertheless, clear hierarchies of space and circulation are shown. The servants' rooms are distributed between the basement and the attic. The basement is reached by separate steps from the main house entrance, so that guests and servants need never see one another. It seems likely that the female servants' bedrooms were furthest from the street, since the maids' and cook's rooms were accommodated in the attic. The grand family rooms, the drawing room, the dining room and the library are reached by a spacious large staircase. The more intimate rooms, family bedrooms and nurseries are reached by a private stair tucked into the centre of the building. The male coachman seems to have had separate quarters and an entrance of his own. It is intriguing that whilst great care was taken to separate servants and household in the parts of the house where servants worked, female servants had to pass through the family's staircase to reach their bedrooms, thus allowing the family to survey and control their movements.

It was the location and ideology of the home which emphasised the gulf between the sexes within the bourgeoisie. Once dwellings became separated from workplaces then it became more difficult for wives to maintain an interest in the husbands' livelihood. This, of course, was reinforced legally by the stipulation that married women should not own property on their own account, a statute which was only repealed in 1882.

The definition of home as a place separate from employment devalues the unpaid work which is done within it, since by definition it is not paid.

This has ramifications for today in that unpaid work has an economic value, estimated at approximately 40% of the value of the country's entire production (Reid & Wormald 1982). The worth of woman's unpaid work also has resonance for council housing. In council housing it is the work which is most like housework, such as cleaning common staircases which has been underfunded and neglected (Roberts 1988).

Finally, the physical separation of home from work restricts the horizons of those confined within it. Previous studies carried out a decade or more ago showed that mothers with young children who were at home all day were more prone to experience depression and isolation than other women or men. A more recent study, however, reversed this finding, probably, the authors thought, because women were able to make a more positive choice for child rearing nowadays (Brown & Harris 1978). Nevertheless, a separation between home and work combined with the absence of transport restricts opportunities for social relations outside the home. In recent years this has become exacerbated by fears of assault at night: a survey made by the Greater London Council when it was in existence found that older women in particular were frightened of going out at night (Greater London Council 1985).

Domestic service

Whilst a celebration of domesticity and an ideology of sexual difference were promoted by an emergent section of the bourgeoisie there existed an important means by which these ideas were made available to working-class women. This was through domestic service. Domestic service provided a significant source of employment for women up until the Second World War. In 1901, 42% of employed women were servants, that is over two million women. By 1931 this figure had declined to 30% of all employed women: domestic service became less attractive as other sources of work became available (Tilly & Scott 1978).

In the seventeenth and eighteenth centuries rural families who could not support their daughters sent them into service with other families. In the nineteenth century this practice continued, but as more people lived in urban areas more servants were in the towns than in the country (Arnold & Burr 1985). Larger households in the mid-nineteenth century had servants to perform specialised functions, such as cook, nursemaid, scullery maid and so on, but in less wealthy households servants undertook a variety of household tasks. Pre-industrialisation servants would have helped with the business or trade of their host household but following the decline of craft manufacture they performed domestic tasks only.

In the nineteenth century the majority of female domestic servants 'lived in'. They were normally young, unless they had some specialised skill such as a cook or a housekeeper, and would leave the household on marriage. After the First World War the number of living-in servants declined and their average age rose. There were still a considerable number of servants who worked on a daily or hourly basis, as a 'char' or daily. Even quite poor households had some kind of outside domestic help: White's account of a model tenement block just before the First World War records how Jewish wives of casual labourers employed charwomen and washerwomen, Jewish and Gentile, in even worse circumstances from the neighbourhood (White 1981).

It is not clear how much of their employers' attitudes to domestic morality were passed on to female domestic servants and how they, in turn were to teach these ideas to their daughters. Davidoff and Hall have described how the mistress–servant relationship was a source of tension in middle-class religious households in Birmingham and Colchester in the early to mid-nineteenth century (Davidoff & Hall 1987). Certainly servants would have been trained in rituals of domesticity. Davidoff has shown how many of these rituals had a tenuous relationship to the efficient performance of household tasks. For example, the dictum that all saucepans should be cleaned every night when it was easier to leave them soaking and clean them the next day suggests that a higher value was placed on tidiness than labour-saving (Davidoff 1976).

Thus, domestic service provided a means by which the ideas and domestic practices of middle-class women could be made available to working women. Furthermore, as the major source of employment for women it supported the equation between women and domesticity.

Morality, employment and housing

This ideology of domesticity pervaded liberal employment reforms. It was not the conditions of women in the sweated trades which attracted the concern of Victorian philanthropists such as Lord Shaftesbury. Rather it was the fate of women miners which aroused bourgeois censure. Pressure brought to bear by an 'unholy alliance' of middle-class philanthropists and working-class male trade unionists led to the passing of the Mines Act 1842, which prevented women from doing underground work. As Humphries explains it was not the apparent dangers of the work which worried the legislators but implications of sexual promiscuity, which they saw as being inherent in women crawling through tunnels, half-naked and sweating (Humphries 1981).

A similar moral anxiety underpinned Lord Shaftesbury's activities in the field of housing. There was much to be concerned about in Victorian working-class housing – problems of supply, rents, facilities, disrepair and overcrowding. It was the latter, overcrowding, which became the subject of the first Parliamentary Act to regulate housing for the working classes.

The Common Lodging Houses Act 1851 regulated the numbers of people who could be housed in any one dwelling at any time. The term lodging house did not mean boardinghouse but referred to lodgings, that is rented housing. The condition of the poorest labourers was such that they could not afford even one or two rooms to rent, but instead had to make do with part of a room. Reformers were aghast at the conditions in which the poorest survived, with men sharing rooms with women and whole families with single men and women. There was a fear of immorality and incest. Evans, in a survey of housing literature of this period, noted:

> Housing literature is shot through with innuendoes and inferences about the perils of intimacy involving men and women, parents and children, youths and the elderly, lodgers and relations, friends and strangers in beds, on the floor, under beds, in every conceivable conjunction and combination. Investigators could reveal grotesque instances of overcrowding but were as much concerned with the moral implications of flesh pressed against flesh as with the more obvious discomforts of piling too many bodies into a confined space.
> (Evans 1978:30)

The Common Lodging Houses Act was strengthened to become enforceable by police inspection in 1853. It is notable that whilst the nineteenth century is often characterised generally as a period of *laissez-faire*, in the field of housing it was a period when many controls were introduced. Apart from the statutes dealing with overcrowding, the bulk of legislation dealt with controls regulating house building. The laws, which were framed as local by-laws set out minimum standards for the construction of houses, drainage, water supply, their ventilation and lighting. Certain of these by-laws governed the shape and planning of houses, by defining minimum distances which houses should be apart in the street and minimum distances between back extensions. Although there were regional differences between by-laws the Model By-Laws of 1877 set out regulations for urban areas and the Rural Model By-Laws of 1901 set out a more relaxed version for country areas.

The by-law house

The majority of the houses built between 1801 and 1911 were terraced houses intended for a single family. Although they varied in size and status they had very similar plans. This similarity, combined with their adherence to the by-laws, which in the latter part of the century were strictly enforced, has led to them being called by-law houses.

These houses were built by private builders, either acting as developers themselves, or for other developers or individuals who wished to invest. The land was sold in plots from larger holdings. Strict stipulations were normally made in the covenant attached to each leasehold, or less frequently, freehold, as to the type of house to be built on each plot and the manner of its rental.

By making these covenants the owners of land were able to ensure, or at least hoped to ensure, that the social class of the eventual tenants could be defined and restricted. Many covenants prohibited particular trades from being carried out within the dwelling. Muthesius cites a Bedford Estates repair lease which in the 1860s outlawed over sixty trades (Muthesius 1982). By defining in the covenant clear specifications for the construction of the house – its decoration, the material and workmanship to be used, the distance of the front garden from the building line and the rents to be set – the land owner could determine the character of an area.

Unlike continental cities, British towns and cities had areas which were segregated by class. In Paris, for example, even in upper-class districts poorer people would have inhabited the attic floors of the five storey blocks of flats. In Britain, such was the segregation between areas and streets that walls were sometimes built in the twentieth century (Tucker 1966). The upper-middle classes were the first to move from the town centres to suburbs in the early nineteenth century: from the mid-nineteenth century onwards developers started building terraces for the lower-middle class and the upper-working class on the edges of towns.

At first it was only those households with a steady wage earner who could afford to live that far from places of employment. Casual labourers, male and female had to live near to their places of likely employment to ensure that that they were available should an opportunity arise. Although it was at first clerks' and foremen's families who moved out to the periphery, later artisans' families followed them. State support was given to this trend through the provision of 'workmen's fares' in the Cheap Trains Act 1883. From the 1860s onwards some London stations had provided a system of cheap fares for commuters and the Act consolidated this nationwide. The fares were about a quarter of a normal fare but class segregation was ensured by cheaper tickets only being valid for trains early in the morning.

One effect of this suburbanisation which took off in the last quarter of the nineteenth century was to strengthen the notion that an ideal family consisted of a male wage earner with a dependent wife and children. Rents of whole terraced houses were such that a single woman, with or without dependants could not afford them. Furthermore, their geographical locations, away from places of casual employment, combined with restrictive covenants against homeworking meant that it would have been difficult for women to earn a wage. Even taking in lodgers, a former

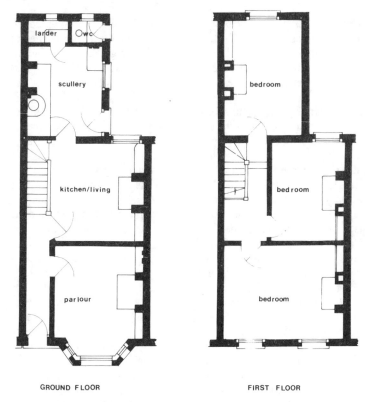

GROUND FLOOR   FIRST FLOOR

Figure 2.2 Plan of typical by-law house

standby of the genteel, was not always possible as some leases prohibited sub-letting. Artisans enjoyed the 'respectability' that a non-wage-earning wife could give them.

    The 'family wage' was an ideal pursued by male trade unionists in the latter part of the nineteenth century. A family wage meant and still means a wage which would be large enough for a male breadwinner to support a dependent wife and children. As Land points out, it is not clear whether working-class militants steadfastly believed in the concept, or whether they marshalled the idea in arguments with employers and government commissions for higher wages because they knew it would be sympathetically received. Nevertheless, it was clear that working men would gain through such arrangements, by a rise in home comforts if nothing else (Land 1980). A non-working wife would be able to give her full attention to the running of the household and when he returned, the husband would have pride of place.

Although the occupation of a single house by a whole family was the ideal, this could not always be achieved. As towns grew, and from 1900 on municipalised trams and tubes pushed the suburbs ever outwards once respectable areas became engulfed. Rents on these houses were such that households had to 'double up' with two in one house. Pember Reeves describes the misery such arrangements caused women, with access to only one wash house and back yard (Pember Reeves 1913).

However, although class differences may have been maintained through rents, location and external facades and decoration, terraced houses had remarkably similar plan forms. There were regional variations and obviously variations due to class in terms of overall size, such as width, height and number of storeys. The essential plan form of the by-law house was standard, with two main rooms on each floor and a staircase at the side, normally leading from a hallway. The back extension or addition, which was built at the same time had a varying number of smaller rooms and floors (see Figure 2.2).

The use of the by-law house combined with its design shows a concern for the appearances of domesticity but with little regard for the conditions in which housework was carried out in the late nineteenth and early twentieth centuries. There was a clear division between the front half of the house on the ground floor and the rear. The facade of the house, its style and ornamentation had been fixed by the lease. Similarly the size of front garden, the type of fence and gate would also have been fixed. The aesthetic unity having been ensured, life in the street was also regulated by local by-laws which prohibited certain types of street games, selling and noisy gatherings (Daunton 1983). The front room or parlour, which faced onto the street, had more elaborate mouldings and fireplaces and was used for receiving strangers, formal family occasions and to keep the best furniture.

The kitchen/living room in the cheaper terraced houses was used for eating in, sewing, looking after children and other domestic tasks. The coal-fired range, which was also used for cooking, was also there. The rear extension, in which the rooms were of meaner proportions, with no architectural decorations or embellishments, was used for heavy domestic tasks such as washing clothes and dishes and cooking utensils, ironing and so on. The back extension might also house a privy and a coal store. The back extension was usually darker than the main house, since it only had light from the side, which was overshadowed by the adjacent house.

By the end of the nineteenth century, by-law houses accommodated thousands of families with lower-middle-class or artisan heads of households. In symbolic terms, as has been argued in relation to the house plans, an emphasis on formality and order to the public side, combined with meanness and squalor to the private side emphasised the subordination of women to men.

This subordination has been 'read off' the design of the house and extrapolated from what we know about the circumstances of working-class wives. Yet there were also positive features of by-law terrace houses. The parlour permitted a degree of comfort in permitting a separation between leisure and housework. Indeed, women who moved from terraced houses to 'open plan' housing in the New Towns after the Second World War regretted the loss of a parlour (Attfield 1989). Furthermore the sheer density of the houses permitted social contact between women over the back fence.

The by-law house had evolved into a peculiarly English house type. Although it was the dominant house type of the nineteenth century, working-class dwellings did take other forms. Inner cities had appalling slums which aroused some reformers to propose new types of housing to solve the problem. It was the activities of the philanthropic reformers which brought flats to some English and Welsh cities for the first time.

Model lives and model dwellings

Philanthropic housing societies were set up from 1847, when the Brideshead Dock Company built what were probably the first working-class flats in England. From then till 1890 housing societies were formed with the aim of solving the housing problem. The societies were founded by well-meaning individuals who raised private funds to build new houses and promised investors a 5% return on their money. The larger of the trusts such as the Peabody, survived into the twentieth century. In 1867 the societies were given state support as the 1866 Labouring Classes Dwelling Houses Act allowed local authorities and housing trusts or societies to borrow from the Public Works Loan Commissioners money to build houses provided that the money was paid back in less than forty years at less than 4% interest rate.

The housing societies did not merely have aspirations to house people. They wished to ensure standards of decency and morality in their dwellings, both reforming the wayward and providing an example for others. The Peabody Trust, for example, had regulations which prohibited the keeping of dogs, hanging out washing, painting or papering their rooms, hanging pictures on their walls or allowing their children to play in the corridors or stairs. At 11 p.m. the outside door to the block was locked and the gas turned out to encourage tenants to go to bed early (Wohl 1977).

Rents for Peabody dwellings were relatively high. In 1875 the average rent for a Peabody flat was 3s 11d a week. The average income of a Peabody tenant household was £1 3s 5d. This, however, was higher than the average wage earned by a male labourer in casual employment, which was 15s. Some single women and other male workers such as costermongers earned less than 10s a week. A representative from Peabody told a Government Committee on working-class housing in 1881 that the Trust only housed the deserving poor and that there existed a class below, a residuum, who could not be housed by the Trust, for fear of dragging them down (Stedman Jones 1971). By 'deserving poor' the Trustees meant men that were in regular employment, who did not drink and who had legal families and women who were clean, hard-working wives who looked after their children and husbands.

The Peabody was not one of the most expensive of the housing societies in terms of rents. The Metropolitan Association, for example, in the same period charged between 8s and 9s a week for a three-roomed flat with a scullery (Wohl 1977). The rents charged varied with the size and facilities offered within the dwellings.

Architecturally, the model housing societies broke new ground. Although a few developments were built which followed the street pattern and design of the by-law house, the most notable example being Noel Park estate in North London, the majority of developments were blocks of flats. Prior to this, high flats had been common in buildings of up to six storeys in Scotland and in two-storey terraced houses on Tyneside, but had not been a common dwelling form in the rest of Britain.

Flats were built because they were the most economical method of providing accommodation at affordable rents in inner-city areas where land costs were high. At first, some societies experimented with building lodging houses as well as family dwellings, but found that family dwellings were easier to manage and provided a better rate of return. In this instance, patriarchal relations and capitalist interests were mutually supportive.

There was some experimentation with the degree of self-containment of individual flats – whereas the Prince Consort's model 'cottages' of eight flats built for the Great Exhibition were completely self-contained, Peabody dwellings shared sculleries and lavatories for reasons of economy. In this aspect of model housing there was a conflict between middle-class philanthropists' desire for separation and social order which would mirror that obtained in upper-middle-class houses and their wish to achieve a minimum of 4% rate of return on money invested. The idea that 'congregation always rears denigration' was strong amongst housing reformers. This conflict was debated at a professional level: for example, when Robert Kerr read a paper at the Royal Institution of British Architects arguing for the construction of single-room dwellings he caused a stir (Evans 1978). Ideological justifications for sharing were also made, one argument being that by having shared facilities, it was easier for the Society as landlord to supervise sanitary arrangements!

Model dwellings, in contrast to by-law terraces were severe and restricted in their external and internal appearance, with a lack of ornamentation and a use of strong, but rather grim materials. As such, they were regarded with a certain amount of dislike by working-class organisations. *Justice*, the newspaper of the Socialist Democratic Federation, called them 'inhuman family packing cases' (Wohl 1977:166) and the London Trades Council attacked the dwellings, referring to them as 'cold, cheerless, uninviting' (Stedman Jones 1971:187).

By the end of the century it had become clear that the philanthropic housing movement could not provide a solution to the problems of slums. This was because the scale of the problem was far larger than that which the associations could tackle, there being insufficient private funding. The inference that only some form of intervention by the state would solve the problem was becoming stronger. However, model dwellings provided prototypes for both policy and architectural form. By choosing to house families rather than single people and enforcing standards of respectability and sobriety and thrift the superior position of the male wage earner as head of household was reinforced. The type of work which women could undertake in order to supplement their income was prohibited, such as taking in washing and lodgers. Architecturally, characteristics of grimness and pared-down finishes as well as the dwelling form of flats were established. Local authority housing followed these examples of social organisation and aesthetic sensibility far into the next century.

Garden Cities

The philanthropic housing movement operated within the structure of the existing city, mainly building high flats in inner areas, occasionally building 'cottage estates' on what were then the outskirts. More radical experiments were made by benevolent entrepreneurs such as Lever and Cadbury, who built entire villages for the workforce of their factories which included communal facilities such as village halls and shops.

Ebenezer Howard's book *Garden Cities of Tomorrow* proposed a new kind of settlement. A public utility company would buy up an area of agricultural land, develop part of it and let the remainder out to individuals and companies to develop. The profit on the increase in value of the land would return to the company through rents and leases: this profit could be used to alleviate the rates and develop facilities such as parks and libraries. The development company would ensure that a green belt remained around the town so that the city did not sprawl. The layout of the town would also be controlled so that factory areas would be segregated from residential areas and a central, civic area could be preserved.

Howard's book was published in 1898 and the first garden city, Letchworth, was founded in 1903. Howard had intended his garden city to be 'town–country' and hence developed at a much lower density than that

FRONT ELEVATION

GROUND FLOOR   **FIRST FLOOR**

Figure 2.3   Plan of Type 'B' house, *Housing Manual 1919*

permitted with by-law housing. In Liverpool, for example, by-law houses were constructed at 31 houses to the acre. Unwin and Parker, architects who supported the garden-city ideal recommended densities of 12 houses to the acre in urban areas. Unwin and Parker were opposed to many of the practices of speculative builders in the construction of by-law houses – in particular the use of ornamentation, a hierarchy of rooms and narrow frontages.

Unwin and Parker drew on the ideas of the Arts and Crafts movement, which proposed simplicity and honesty of materials and construction in order to achieve elegance and beauty. They were also opposed to the artifice of designing houses with one formal room for public show and smaller family rooms and sculleries at the rear. Parker commented on such plans:

> In the greater number of these houses the third room is never used, or used merely because it happens to be there, and its chief end seems to be to provide a place for the women of the household to spend any spare time they may have, cleaning down, dusting.
> (Unwin & Parker 1901:3)

Instead of narrow-fronted, decorated houses with gloomy rear extensions Unwin and Parker developed designs for wide-fronted cottages with rear extensions consisting only of a scullery and wash house and large kitchen–living rooms and good-sized bedrooms (see Figure 2.3). They wished to make the kitchen–living room a family room in which all manner of tasks could be undertaken: they even designed plate racks which fitted in with the overall aesthetic of the room.[1] Their ideas, which were radical, met with some working-class resistance however. Swenarton reports that houses at Letchworth designed by them without parlours were strongly objected to (Swenarton 1981).

The Garden City Movement gained in strength in the first decade of the twentieth century with the construction of Letchworth and a number of garden suburbs, the most notable being Hampstead Garden Suburb. Even the houses within the garden cities and suburbs which were aimed at working-class tenants were more expensive to rent than by-law houses, because of the increased plot size and the size of the house itself. The layout of the garden city, with its clear separation between work and home and its single-family housing let at a rent which only a skilled manual worker could afford, reinforced trends of women's dependency on men and the cultivation of an ideal family of a working man with dependent wife and child. The internal layout of the houses introduced a new way of regarding domestic work by dignifying it as part of family living rather than designing spaces so that all was kept from public view.

Although the main trend of garden-city planning was to encourage respectable artisans to support a dependent family, alternative domestic arrangements were the subject of more middle-class experiments within two garden cities. These experiments in co-operative living were radical, but nevertheless should be seen in the context of their revolutionary antecedents which occurred over half a century before.

Co-operation and socialism[2]

The processes which have been described so far have involved segregation and stratification, between the sexes, between classes, between city and suburb and between spaces within the house. These demarcations were evolving within a social and aesthetic framework which was aiming towards order and unity, for example, within the appearance of the street or through the operation of model building by-laws. Whilst these processes might be termed dominant in Williams' (1973) use of the word, there were other trends or movements which were in opposition.

Taylor has provided an account of the early Owenite movement in Britain in which ideas about women's equality as well as equality between men vied for importance. In order to found a new way of living, a break from the old immoral world, these early socialist–feminists founded seven colonies or communities, where they could live and work (Taylor 1983).

The colonies were short-lived, suffering from economic problems as well as the crises of leadership. Despite attempts to minimise gender differences with agreements about 'free unions' or free love and attempts to socialise domestic work with cooking rotas and nurseries there was still considerable friction between the sexes. Women performed the majority of domestic tasks as well as working for the commune. Indeed in the last days of the biggest colony, Queenwood or Harmony, women helped to keep it going with heavy manual labour. The communes collapsed for a variety of reasons – lack of funds, the tensions brought about by sexual libertarianism and the financial extravagance of Owen himself.

Although such radical experiments were not to be repeated again until a century later, ideas about changing the nature of women's domestic work by arranging for it to be done communally provided an undercurrent to housing reform. Unwin and Parker had proposed a 'co-operative quadrangle' at New Earswick which would have had a communal kitchen. They justified this on the basis of economy in terms of labour and in improvement in quality of service.

> Instead of thirty or forty housewives preparing thirty or forty little
> scrap dinners, heating a like number of ovens, boiling thrice the
> number of pans and cleaning them all up again, two or three of them

retained as cooks by the little settlement would do the whole, and could give better and cheaper meals into the bargain.

(Unwin & Parker 1901:104)

The quadrangle at New Earswick was never built. Other co-operative dining schemes were – at Letchworth. Two schemes were built, Homesgarth and Meadoway Green. Homesgarth was for middle-class childless couples and was based on a courtyard of flats which did not have kitchens. There was a communal dining room and laundry and all domestic work was done by servants. Meadoway Green provided for a more working-class tenancy – each of the dwellings had small kitchens, but the courtyard as a whole had a dining room large enough to seat all of the tenants, plus guests. Women residents took it in turn to manage the catering. A full-time cook and a part-time charwoman were employed to do the domestic work itself. Thus neither existing-class relations nor indeed the sexual division of domestic labour were challenged (Hayden 1981).

These experiments were small-scale and it was difficult to know how widely they were publicised. In one way they were drawing on earlier working-class arrangements which had been made through poverty: in some villages, for example, there had been a communal bakehouse simply because people could not afford ovens. In this way they may have contradicted the wishes of the wives of skilled workers for signs of greater prosperity and autonomy. Macfarlane reports that when in 1918 the Women's Sub-Committee of the Tudor Walters Committee investigated 'working women's' wishes for their housing, they found that communal dining rooms did not meet with an enthusiastic response (MacFarlane 1984).

This tension between an idea of stigma attached to sharing and a notion of a respectability that comes from privacy and self-sufficiency is a theme which runs through nineteenth- and twentieth-century housing design and policy. It may be seen at many levels – for example, in model housing between providing self-contained accommodation and flats with shared WCs and sinks, in council housing a tension between providing shared laundries and individual washing machines and in speculative housing between communal green areas and individual gardens. As has been seen, nineteenth-century ideas about housing were concerned with values of decency and privacy: these ideas were to be challenged in the early twentieth century by an architectural ideology which sought to sweep away prejudice with the clean broom of rational thought.

The new objectivity and housework

Whilst exponents of the English Arts and Crafts movement, such as Unwin, emphasised honesty of expression and truth in materials, they did not break the boundaries of traditional construction. Furthermore their use of traditional building materials and features such as decorative tiling and timber panelling has an almost whimsical quality. In Germany and Austria, in cities which also faced the problem of providing cheap housing for working people, different kinds of avenues were explored.

Architects such as Mart Stam and Hannes Meyer had evolved a different theory of architecture in which social and aesthetic goals were also intertwined, just as Parker and Unwin's were. They wanted architecture to do away with the decoration and pomp of the nineteenth century and in a sense to become classless – a more materialist expression of construction and the function of the building. They also wanted architecture to address the problems of twentieth-century living, in particular mass housing, which some architects did not regard as 'architecture' in the sense of art.

There is much to be said about the work of these architects. The aspect of their work which is relevant to this book lies in their espousal of rationality or objectivity (*sachlichkeit*) and the way in which they applied this to the design of kitchens and the use of domestic appliances. The new objectivity which these architects espoused was an optimistic use of twentieth-century technology combined with detailed attention to the use or function of the building. Stam and Meyer in particular were in ideological sympathy with the early ideals of the Russian revolution of 1917. Their aesthetic ideas had influence on and were influenced by leading architects of the period such as Gropius and May.

Flats which were designed by this grouping were built with an aim of providing a high standard of services such as hot water and heating at a low rent. Frampton reports that May achieved this in large estates of housing around Frankfurt largely through the use of the Frankfurter kitchen, which was designed by Schutte Lihotzky in 1925 (Frampton 1985). This kitchen, which was based on the galleys used on international trains, made a maximum use of equipment in a small area. This kitchen was regarded as 'efficient' because it made an efficient use of space and because the person working in it made the minimum of movements. Thus a high standard of provision could be made at a low cost. Other architects developed the idea of providing tightly packed but highly serviced kitchens, designing them not for working-class houses but for large villas for the middle classes too.

At the same time as architects were designing kitchens for maximum efficiency in spatial terms, Christine Frederick in the US was also redesigning kitchens, but to increase the efficiency of the woman working in them. Frederick used the technique of time and motion studies to plot the optimum position of workbenches, sinks and cookers (Hayden 1981).

Both the ideas of Frederick and of progressive architects helped to reduce some of the effort needed for housework: electricity and running hot water being of paramount importance and planning of some minor significance. However, their efforts did not address the central problem of efficiency in that housework, as an unpaid task, is not amenable to rationalisation on the lines of factory production (Ehrenreich & English 1979).

However, the ideas of efficiency in spatial terms and ergonomic efficiency, combined with the use of modern services became part of the Modern Movement in architecture's method of kitchen design. This led proponents of Modern Movement architecture, such as F.R.S.Yorke, who wanted to introduce the style into Britain, to claim it as labour saving and therefore liberating for women (Yorke 1934). In this way post-war architects saw themselves as being sympathetic to feminism, because they espoused modern technologies such as central heating, despite other faults in their buildings.

The design of workers' flats in Berlin, Frankfurt and Vienna by architects who were committed socialists also led to an association between flats and socialism. This attitude will be illustrated in a later chapter where a Conservative Member of Parliament is reported as making this connection. Whilst European architects designed flats in keeping with their traditions of city planning, the English socialist tradition of William Morris and Unwin was of small houses in a rural setting. Thus some important technical developments became value laden to an inflated degree.

The mainstream of English domestic architecture had been evolving in a single-family-house tradition, based on high rents and low densities. By contrast European socially committed architects had concentrated on low-rent, high-density dwellings. Some flats were built by local authorities in Britain in the inter-war period, but these were expressly subsidised for slum clearance at low rents and with low standards of provision.

Suburban housing

The First World War was to prove a watershed in terms of state housing policy. There had been a housing shortage before the war began; this was exacerbated by the lack of construction during the war. Due to the actions of ordinary women in Scotland who went on rent strike in protest at rent rises an emergency Act had been passed in 1915 which fixed rents at below a free market level (Orbach 1977). Private capital therefore had less reason to invest in speculative housing. There were a number of serious strikes in 1918 after the war ended: these included a strike by the police. The political situation was thus extremely volatile, and of grave concern to the Cabinet since the men returning from the war knew how to use arms (Marwick 1970).

Expectations had been raised during the war of better housing for the working classes. A government committee, headed by Tudor Walters, had investigated the design and provision of working-class housing. The Report produced by the Committee had come down firmly in favour of Unwin and the Garden City Movement's ideas for lower densities, wider frontages and a reduced number of rooms. The Report was also somewhat cautiously in favour of higher standards of fixtures and fittings than had previously been provided in working-class dwellings and so recommended the provision of a separate bathroom, a kitchen range, a copper to heat water, a sink and relatively generous space standards.

1919 saw further labour unrest and, in response, the Addison Act was introduced which required local authorities to make a survey of housing needs and to build council houses subsidised by the product of a one-penny rate. The houses were to be built on the design guide lines laid down in the Tudor Walters' Report. The rents for the houses were relatively high and as one MP remarked, it would only be the 'aristocrats' among working men who could afford them (Swenarton 1981:83). The reasoning behind providing a small number of high-rent, high-standard houses rather than a large number of low-rent, lower-standard houses was concerned with ideas of filtration and obsolescence. If the better-off sections of the working class moved into council houses, then this would free good-quality privately rented housing for those who could afford it and create further vacancies in poorer accommodation. If the council houses were not built to a high standard, then working-class aspirations would rise and the dwellings would become obsolete before their forty-year loan period was up. Although the latter aspect of this reasoning may have been correct, it had the effect of making subsidised housing only available to skilled workers and their families.

Since the new council houses were built at a low density they were of necessity at the peripheries of built-up areas. Consequently, it was not only households with skilled male workers at their head who could move into them, but skilled male workers who did not require their wives to provide an additional source of income, since the houses would have been built at a distance from places of women's work. Sub-letting was prohibited, to avoid overcrowding.

Local authority building programmes under the 'Addison' Act quickly ran into problems with rising building costs and shortages of skilled labour. The Treasury was against public expenditure on housing and in 1921, when labour unrest appeared to have died down, the programme was curtailed. Local authorities were again allowed to build under the Wheatley Act of 1924 which was codified by a subsequent Conservative administration. Space and standards of fittings were eroded, however, from their high point in 1919 as building costs rose, although a minimum standard was enforced.

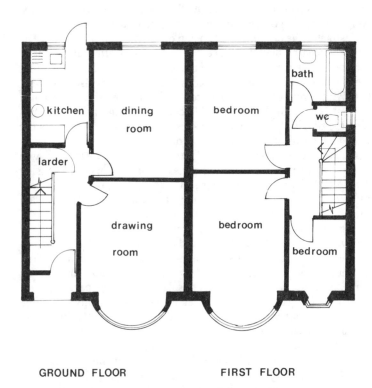

GROUND FLOOR   FIRST FLOOR

Figure 2.4   Plan of typical suburban semi-detached house

The Addison Act itself and the Tudor Walters' Report were of great significance, not only in setting design standards for council housing, but in setting standards for speculative building. In order to attract buyers speculative builders had to make sure that their houses were equivalent to those offered by the state. Between 1918 and 1934 1.75 million houses were built by private developers (Bowley 1945). The majority of these were semi-detached houses built in the suburbs.

Private house building was encouraged by central government with the introduction of the Chamberlain Act in 1923. In this Act private developers were given grants to construct houses for either sale or rent, provided that the houses were built between certain maximum and minimum dimensions. Local authorities could also make grants to prospective purchasers for deposits or repairs. In this way owner occupation was encouraged by central government, a policy which was to continue, although not with direct subsidies, throughout the inter-war years. Building societies, which had been regulated by central government since 1874, provided mortgages. Although some houses were built for rent, the overwhelming majority were built for sale.

At first, speculative builders aimed to attract buyers from the middle classes – professional and white-collar employees. However, after building costs dropped in the 1930s, builders were able to extend their market into the lower-middle classes and even into the higher-paid working class. Builders arranged attractive loans with building societies with small deposits – even as low as £5. For households living in two rooms in the inner city, with no electricity and a night-soil bucket, the incentive to save and get out was high (White 1986).

A typical plan for a speculatively built inter-war, semi-detached house was a combination of by-law house formality and Unwin's concern with sunlight and fresh air (see Figure 2.4). The Type C plan recommended in the Tudor Walters' Report became the standard plan for builders. It had a parlour at the front, a living room at the back with a range, a small kitchen alongside the back living room and a hall with stairs going up to the first floor. The first floor had two bedrooms, a minimal bathroom and a small bedroom, which was also known as a 'box' room.

Unlike by-law houses, semi-detached houses were built with rudimentary damp-proof courses, electricity and.fitted bathrooms. Although they were heated by coal fires and back boilers, running hot water was available for the kitchen sink, the bath and the wash basin. Often builders offered the lure of a domestic appliance, such as a refrigerator or a vacuum cleaner as an incentive to buy.

Women were targeted by advertisements for the sale of these houses as housewives and homemakers. The labour-saving properties of the houses were emphasised, as were their low cost and cleanliness. Whatever reductions were made on construction or space standards at the lower end of the speculative market, builders were always careful to provide the most up-to-date services such as electricity, hot-water heating and coal stoves. This tendency led Bowley, herself a woman, to comment on the growth of owner occupation 1918–1939:

> It marked the first stage in the battle to break free from the pattern
> of housing conditions and domestic life created in the Nineteenth
> Century; in all senses it was a logical part of the emancipation of
> women, an attempt to free themselves from household drudgery.
>
> (Bowley 1945:83)

Demographic factors encouraged a surge in owner occupation. The age of first marriage dropped, which meant that more households were formed. The birth rate also dropped, through a combination of greater sophistication about birth control and a desire by women to have higher living standards by limiting the size of their family. This meant that a three-bedroomed house could accommodate a household of up to five persons and moreover there would be surplus income to provide carpets, furniture and vacuum cleaners.

State support of owner occupation as a form of tenure clearly provided the social underpinning for men to be the head of the household. Women were not regarded by building societies as suitable mortgagees until the passing of the Sex Discrimination Act in 1967. This is not to say that no woman ever got a mortgage before then, but that it was extremely difficult for a single woman to be given a loan without a male guarantor. It was even difficult for married women to have mortgages in their own right (Gittins 1985). Furthermore, even the cheapest house required a regular substantial income to repay the loan advanced on it.

Not only was men's superiority reinforced as a provider of shelter, but the location of speculatively built houses reinforced women's secondary position in the labour market. In the suburbs transport would be provided for men to get to their work in the city centre or local places of employment. Women were far from the small-scale manufacturing industries which traditionally had provided employment. Where factories did relocate, or new ventures started which employed women, such as electronics factories, then they were able to draw on a pool of female workers who had few other opportunities for employment. Lack of suitable employment became an embarrassment for the London County Council when, having built a large suburban estate in Becontree in East London, they were forced to advertise for tenants on the side of buses, despite their overall housing shortage (Jackson 1973). Although it was lack of male employment which caused problems at Becontree, it is possible that lack of female employment may have deterred other households from moving to the suburbs.

Certainly for households on a low income, suburban living might have meant higher housing standards rather than higher living standards. In a famous study of Stockton on Tees, the medical officer of health found that death rates were higher on suburban council estates than in the town because food and transport costs were higher (Mackintosh 1952). Prior to this, suburban living had been thought to be healthy because of the cleaner, less-polluted air (until the 1957 Clean Air Act town centres suffered from atmospheric pollution) and because the houses themselves had electricity there was less coal dust within them.

The architectural profession, however, despised suburban housing. It was criticised for being monotonous and lifeless. This prejudice against suburbia continued after the Second World War, up to the 1970s. An architect author of a book celebrating the 'suburban semi' describing his suburban upbringing met with contempt from his tutors during his architectural education (Oliver, Davis & Bentley 1981).

Concluding comments

The evolution of British housing from the early nineteenth century onwards was strongly influenced by an ideology of family and respectability in which notions of class and gender were closely intertwined. From the 1780s onwards the middle classes gradually began to differentiate themselves from the aristocracy and from the working class through their entrepreneurial activities, religious morality and cult of domesticity. Middle-class women became closely connected with the home and their opportunities for advancement in the public realm became severely circumscribed.

Speculative builders responded to the immense demand for housing by building, in a peculiarly English form, by-law houses which at the same time provided relatively high-density dwellings per acre and satisfied the aspirations of a strata of skilled working-class people. By-law houses were built on the peripheries of towns and cities: as cities grew successive waves of houses were built until what had been the periphery became the city centre. Different strata of middle-class and working-class households were accommodated in different grades of by-law house. These were differentiated by size, external and internal decoration and by restrictive covenants on their use.

Philanthropic responses to the nineteenth-century housing crisis resulted in building a new form of dwelling – large blocks of flats. These, despite being ugly, were administered in a way which supported sobriety and propriety amongst their inhabitants. Single women also had difficulty in gaining access to these because of high rent levels.

Collective working-class action, in combination with other factors, ensured that building rented housing was no longer profitable after the First World War. The construction of council houses promoted a new housing form which was an amalgam of contemporary concerns with sunlight and fresh air, standardisation, domesticity and a general raising of working-class living standards. This new form of housing had been designed with garden cities specifically in mind, with their combination of town and country. Local authorities could not build new towns, so the new council houses were built in suburbs. Like middle-class suburbs a century before them, this type of housing promoted a stable family life through its exclusive nature and distance from other amusements. Unlike middle-class housing women did their own domestic work, albeit with improved equipment.

In the inter-war period the coalition governments reduced the standards of council houses and promoted owner occupation. Owner-occupied houses were built predominantly in the suburbs and in a characteristic form which was derived from council housing although with echoes of the by-law terrace. Women were excluded from the purchase of these houses independently of marriage. The fittings of suburban houses were promoted with women in mind with the inclusion of fatigue-reducing services such as electricity and piped hot water. Such houses were only available to the more well-off sections of male wage earners.

Flats continued to be built by local authorities, but mainly to a low standard of space and services for rehousing slum dwellers. Despite the problems of bringing up children in flats, which began to surface in the 1930s, architects became interested in them as a dwelling form. The flats which had been built by the architects of the 'new objectivity' school were beginning to receive publicity. Architects could look to these European examples as successful models for providing low-cost modern housing.

Thus the mainstream of British housing policy supported the notion of high-class housing being funded by a male wage earner who could securely and comfortably house his dependent family. Whilst the ideal might have been that wives did not engage in either waged employment or heavy household tasks, the reality was that for many households a wife's wage meant the difference between survival and abject poverty.

By the Second World War increasing attention was being paid to the fact that only richer middle-class families had servants and that, therefore, wives had to undertake the burden of domestic work. Experiments to reduce domestic work through the provision of dining rooms had been tried amongst the avant-garde, but without widespread success. Despite the fact that the technology existed to make housing more comfortable with central heating and modern appliances, the distribution of wealth was such that only the relatively well-off could afford this standard of housing. It was with these disparities and contradictions between ideal and reality, supply and demand that Britain entered the Second World War.

Chapter three

Women as homemakers II

After the First World War the government had felt it necessary to introduce state housing, that is housing provided by local authorities for the first time. Once industrial unrest had died down and in the period of coalition government a commitment towards state housing was diluted and owner occupation promoted. Prior to the Second World War, state policies towards housing and planning had been, if not *laissez-faire*, at least premised on the idea that market forces should predominate. During the period of the Second World War and the Labour government which followed it state policies assumed a much more directive role. Indeed, much of the bureaucratic apparatus for implementing the policies after the war had been set up as part of the war effort itself (Cullingworth 1975).

Four major Government reports on planning and housing were produced: one was published just after war broke out (Barlow Report 1940) and three in the course of the war itself (Dudley Report 1944; Scott Report 1942; Uthwatt Report 1942). Two plans were produced for London and the London Region – the County of London Plan 1943 and the Greater London Plan 1945: again both had been published by the end of the war. The Beveridge Report, which was published in 1942, was also of importance in setting out not only a plan for social services after the war, but also providing intellectual and moral justification for such a welfare system.

Although these documents might appear at first sight to be dry and without any reference to gender, they, as well as other published and unpublished memoranda and reports, provide an unusually rich record of the aims and values which underlay a broad range of policies connected with housing. This source material will be examined in the following three chapters in terms of housing form and reproduction, the location of housing and gender divisions and design standards and class-status divisions.

In considering this immediate post-war period it should be remembered that it was a period of hope for the future, that Britain could be reconstructed differently from pre-war ways. Attitudes towards physical planning were welcoming, so much so that an entire edition of the popular magazine *Picture Post* was devoted to planning and housing.[1] In it architects such as Maxwell Fry and Elizabeth Denby wrote optimistic articles about the benefits of new housing and well-planned towns, promising leisure, health and cleanliness. A simplistic attitude towards planning prevailed: social and physical goals were merged in an ill-defined way. The prevailing attitude was characterised by Cullingworth in his history of planning: 'If the war could be "planned", why not the peace?' (Cullingworth 1975:1).

It is this possibility for change which makes this period such an interesting focus for study. As I have indicated, industrial unrest was at a height after the First World War leading the Cabinet to worry, for a short period at least, about the possibility of revolutionary change. However, this was in the context of a Britain which had largely been left physically untouched by the war itself. The situation was different after the Second World War. Nazi bombers had wreaked havoc upon British towns and cities leaving 200,000 houses destroyed and 0.5 million severely damaged or uninhabitable (Gibson & Langstaff 1982). Little house building had taken place during the war. Moreover, whole neighbourhoods, such as for example, Coventry city centre and parts of the East End of London, had been so severely damaged that reconstruction rather than piecemeal redevelopment was a necessity.

Planners and politicians themselves were enthusiastic about the new order which could be created. It was suggested that Hitler's bombers had achieved more in slum clearance in four years than a century of public health legislation had made possible. So great was the enthusiasm of at least one architect for this process that members of the public were offended by his apparent callousness (Ravetz 1980).

This is not to imply that planners and architects were responsible for the policies and designs that were pursued. Housing form, as the previous chapter has shown, is not the subject of an arbitrary whim by architects but is the product of a complexity of social forces in which masculine assumptions and ideas about women's role in the family and at home play a key role. Moreover, reconstruction involved changes in legislation, the introduction of new bureaucracies and controls as well as massive human effort in physical construction. None of these processes were without argument or conflict.

This chapter will examine one small part of this process. I shall look at the anxieties expressed during and after the Second World War about the falling birth rate and discuss these in relation to controversies over whether to build houses or flats in the immediate post-war period.

Family policy

The challenge which the Nazis posed to Britain was not only military, but also one of ideology. A War Aims Committee of the Cabinet had been set up in 1940 to formulate a statement which would be used as counterpropaganda to the Nazi's new order in Europe. The statement was to contain a list of principles on which a new society in Britain would be based. The final draft of the statement included four principles. The first three asserted the importance of religious, moral and social freedoms, of equal opportunity, justice and the rule of law. The last was: 'the *domestic* principle of the sanctity and solidarity of family life' (Cullingworth 1975:4–5, emphasis in original). The statement was not published; nevertheless the wartime Reconstruction Committee of the Cabinet had made a commitment to an ideal of the promotion of family life.

The Beveridge Report, which was published in 1942 was part of this commitment. In the introduction to the Report, Beveridge noted that his proposals for the abolition of want through a system of social insurance was formulated within a recognition of two major demographic trends. The first of these was an increase in the numbers of old people and the second a decline in the birth rate. This last factor, Beveridge argued, should be reversed; otherwise he saw a rapid, increasing decline of the population taking place. He thought that to avert this possibility, primary importance should be given to the promotion of motherhood and to the care of children in the post-war schemes for social expenditure.

In a speech which he made following the publication of the Report, Beveridge stressed the important part which domestic architecture and architects had to play in raising the birth rate and strengthening family life. He thought that architects had a significant responsibility in designing homes which would accommodate the products of an increased birth rate and moreover be cheap to build and easy to run.

> Architects should set out to be the Lord Shaftesburys of the home.
> That means thinking not only of the walls or roof or of the shape
> and size of the rooms, but of every detail of equipment and its
> placing. That means thinking of how to make homes not only well
> but quickly and cheaply. It is important also that those who design
> homes today should realise that they must be birthplaces of the
> Britons of the future – of more Britons than are being born today. If
> the British race is to continue there must be many families of four or
> five children.
>
> (Beveridge 1943:172)

In this speech Beveridge was talking within the climate of pro-natalist opinion which surrounded 1940s discussions of social policy. Alarm had been experienced by both conservative and socialist politicians and campaigners about the decline in the birth rate in the 1930s. This alarm was fuelled by uncertainty: no census was taken in 1941 because of the war.

In fact although the birth rate declined to a low point in 1941, by 1944 the sum total of legitimate and illegitimate births was 847,419, a figure not equalled since 1923 (Riley 1983). Demographers, however, based their projections on pre-war patterns and trends, failing to take into account the lower age of marriage and earlier child bearing. Thus their prognosis of a declining population was not borne out in practice.

'The family' became an object of rhetoric in wartime speeches and propaganda. Because pro-natalism had been espoused in an authoritarian way in Nazi Germany and Fascist Italy measures were proposed which ameliorated the strains of family life, rather than forcing women to have babies. These proposals included a wide range of social provision such as improved obstetric and gynaecological help for working-class women, holidays for tired housewives, day and night nurseries, communal restaurants and laundries, family tickets on trains, after-school play centres and official baby-sitters.

Securing a stable family life had important social and economic implications in terms of the distribution of industry and people, maintaining a thriving agriculture and in housing standards. The Barlow Report, published in 1940, tackled the first of these issues.

In the period between the First World War and the Second World War British industry went though a painful and rather inadequate period of structural adjustment. Whilst the 'traditional' staple industries, iron and steel, ship-building, cotton, enjoyed a small boom after the First World War this was quickly followed by a long period of decline. This decline was caused by adverse trade conditions and an inability to keep pace with technological change. Unemployment in the regions occupied by these industries, that is, the north-east, South Wales, central Scotland, Lancashire and west Yorkshire, rose to 14%. As traditional industry stagnated, so the regions in which they were based declined economically.

Whilst the slump proceeded in these areas, the south-east of England and the west Midlands enjoyed an unprecedented boom. This was based on the expansion of modern 'light' industries in these regions. Examples of 'light' industries were electrical engineering and vehicle and aircraft manufacture. The fuel source for these industries was electricity which meant that they could locate near towns and ports rather than near steel plants and coal mines. Between 1921 and 1937 the rate of increase of population in London and the Home Counties was nearly two and a half times that of the country as a whole. By 1939 London and the Home Counties contained approximately one-quarter of the population of England, Scotland and Wales. Some established industries also grew, for example, building, building materials, tobacco, food processing, furniture manufacture, printing, footwear and hosiery. There was a rapid increase in road transport and a slow decline in railways. Thus a severe regional imbalance was set up in Britain (Glynn and Oxborrow 1976).

London changed in character from a city based on a port and a centre of handicraft production to a factory centre. The main industries were leather tanning and dressing, boot and shoe manufacture, the finishing sections of the iron and steel trades, tailoring, dressmaking, millinery, the production of chemicals, dyestuffs and drugs, ship-repairing, and grain-milling. London was the centre of the timber trade and also of the printing and publishing trades. It had a number of services connected with public institutions and tourism. The activities of the port continued and London was a centre for warehousing and distribution based on the port itself.

The suburbs of London expanded as explained in the previous chapter. Thus there was a severe imbalance between the declining depressed areas with their old towns and cities and terraced houses and the booming south-east with its suburban sprawl. Because suburbia developed where land was cheap and new transport routes were opened, this meant that homes were not necessarily near jobs and many experienced long, fatiguing journeys to work.

The Barlow Commission, whose report was published just after the outbreak of war in 1940, proposed to redress this situation in a variety of ways. It recommended that the concentration of industry in London and the south-east, and to a certain extent in the west Midlands, be dispersed and decentralised by voluntary means. A National Industrial Board should be set up to advise and research on decentralisation, dispersal and development. The Commission recommended that the optimum distribution would be for a balance of industry throughout Britain, and that each area should have a diversity of industries to counteract depression. Reducing the concentration of people in towns and in London particularly was to be achieved by a series of proposals which had been propounded by the Garden City Movement for over half a century. These were for garden cities on green field sites, satellite towns near the major conurbations, trading estates, town extension schemes and other small-scale projects. The Report proposed compulsory powers for the National Industrial Board to prevent any further growth of London and the Home Counties.

Although the Commission considered the economic and strategic reasons against concentrations of population in towns and cities it also examined the public health arguments about the effects on health of living in large conurbations. The Commission was not wholly given over to the view that cities were unhealthy in themselves. London, their report recorded, had had a lower death rate than the national average in the years 1921–1930. Overcrowding, lack of planning, smoke, noise and the economic state of families, they suggested, were all important factors in the health of towns.

Furthermore, the Registrar General had been keen to urge the Commission not to substitute higher standards of housing for a decline in nutrition, noting that when food was short, it tended to be married women and children who suffered particularly. This was because in situations of poverty, either caused by the husband not giving his wife adequate housekeeping money or through low wages, married women tended to prioritise their husbands' food. Often wives gave their husbands the best share, then the children and failed to eat enough themselves (Spring Rice 1939; Delphy 1974).

The Commission thought that towns could be planned for healthy living provided that there was slum clearance, better drainage, more sunlight and open space. A figure of 6 acres of open space per 1,000 population was recommended. The close proximity of residential areas to heavy industry with smoke and noise was reported as being detrimental to health.

In commenting on the type of housing to be provided the Commission noted the unpopularity of flats. Concerns were voiced for the health of children, given the noise problems and confinement of flats and also the expense of constructing them. It was also noted that working men liked the privacy of a house and garden. Women's views were not recorded. It is a staggering indictment of pre-war attitudes that these crucial considerations of the domestic sphere should have been made without a clear representation of the views of differing women's organisations. There was only one woman on the committee, Hermione Hichins and she contributed to a minority report which took issue with the recommendations of the main committee.

Although a TUC official was recorded as having an opposing view, the Commission concluded that a certain amount of flat building was inevitable in towns, since land values were high there. The Commission did not favour suburban housing either. It found that community life was made more difficult: also that women had to pay higher prices for food and other basic necessities than in the old inner-city areas.

The Commission concluded that the most healthy conditions for the future of the labour force lay in replanned towns and new 'balanced' communities, with plenty of sunlight and fresh air. A minority report, written by Abercrombie, Elwin and Hichins, considered that the main report had not emphasised sufficiently the perils to health of congested towns. They warned of the effects of living in flats in lowering the birth rate:

> Further the opinion is widely held that the rapidly proceeding
> transfer of a large part of the urban population from houses to small
> flats in which they are 'cabin'd, cribb'd, confin'd' with scarcely any
> room to spare, means that any increase in the present size of family
> would present great difficulties in regard to accommodation. When

we recall that the net reproduction rate in this country is at present 0.75 the importance of this question is obvious.

(Barlow Report 1940:225)

The minority report urged greater controls on the location of industry otherwise 'the costly flat and fringe re-housing policy of the Government' would cause more problems.

The Barlow Commission recognised that in proposing a compulsory system of planning controls the question of development rights over land would have to be considered. If the state were to compulsorily purchase land, then compensation would have to be given: conversely, if the state were to initiate development, then private land owners might benefit from an overall increase in land values, in which case some tax on the increase or betterment might be due. Obviously powerful vested interests were involved: in order to consider the matter further the 'Expert Committee on Compensation and Betterment' otherwise known as the Uthwatt Committee was brought together and its final report was produced in 1942.

The Uthwatt Committee stressed the importance of rebuilding towns and cities in order to ensure 'health and efficiency' (Uthwatt Report 1942:17). They argued that this was one of the most important tasks for the post-war world. The measures which the committee proposed for physical reconstruction included the provision of open space and rebuilding congested areas. These areas were to be rebuilt at reduced densities from the existing if necessary and replanned with a separation of industrial and residential spaces, although the distance between them was not to be so great as to necessitate long journeys to work.

The committee emphasised the deleterious effects of high land values on housing in built-up areas in that it meant that flats rather than houses were built for workers and encouraged overcrowding. In order to avoid this situation the committee recommended that existing legislation on compensation and betterment be changed. The changes they proposed were far reaching – a vesting in central government of all development rights in land outside of built-up areas; wider powers for local authorities to compulsorily purchase land in built-up areas; a periodic levy on an increase in site values in built-up areas and the setting up of a Central Planning Authority and research team. Since previous legislation, passed in 1932, had only permitted local authorities to make local plans, these proposals were considerable reforms.

The Barlow and Uthwatt committees had considered replanning towns and cities from the point of view of the orderly and healthy reproduction of the workforce. Women had been referred to from a pro-natalist standpoint – that is attention had been focused on raising the birth rate and women had been viewed solely as mothers and not as wives or independent beings. Another wartime report, the Scott Report, published in

1942, which examined the use of land in rural areas took the same attitude towards women, although unlike Uthwatt and Barlow, the committee referred to the problems which the widowed and unmarried faced.

The Scott Report, in examining rural land use, had to consider the housing of agricultural workers and also the demands which the growth of cities made upon land generally. Perhaps because the National Federation of Women's Institutes submitted evidence the Report stated a particular concern about the poor conditions which women suffered in tied housing. They were worried about the isolation of rural tied cottages, of their distance from shops, health centres and schools and of the consequent loneliness which young wives would suffer. If wives were lonely and unhappy, they reasoned, then the education of the children and the health of the whole family might be adversely affected.

Poor rural housing raised the spectre of a lowered birth rate. The committee stated that they had received evidence that couples were limiting their families because of having small houses with not enough bedrooms. The report also noted, progressively, that there were not enough small dwellings available for those falling outside the framework of a nuclear-family household, such as the widowed, the elderly and the unmarried.

The Scott Report recommended that better rural housing should be provided. In order to achieve this, farming should be made more efficient and therefore urban sprawl should be contained. The report argued that if flats were built, with plenty of light and air circulating around the buildings, then the problems of urban sprawl would be diminished. The implication of this argument was that for good agricultural land to be preserved, built-up areas would still have to be developed at high densities, which would necessitate the building of flats.

Thus, securing a stable family life had important social and economic implications in terms of the distribution of industry and people, housing standards and maintaining a thriving agriculture. The Barlow Report had suggested that decentralisation would support strong industry and good housing standards. The Scott Report had warned of the dangers of such decentralisation encroaching upon useful agricultural land. Yet both reports acknowledged the need for good housing conditions, even though the need was couched in pro-natalist terms. To achieve good housing and an even spread of industry meant planning the location of industry, housing and agriculture: that is, state intervention in the free market for land. The Uthwatt Report had recommended that rather than undertake the wholesale nationalisation of land, the state should exercise control over development by a series of measures including the nationalisation of development rights outside built-up areas and taxation on increases in site values.

These reports raised not only complex legal and technical issues but also questions of political principle. In effect a considerable increase in state intervention in free markets was called for, and an unprecedented control of the location of investment. It is not surprising that the Coalition government, whose main purpose was to fight a total war, was not able to take action on these issues.

Scott, Uthwatt and Barlow had reported by the end of 1942 and their recommendations had been made public. Local authorities were then asked to draw up housing programmes for their localities, to be implemented immediately war ended. In the absence of any clear decision from central government as to whether to set a up a central planning authority, or indeed any further planning controls, their task was not easy (Cullingworth 1975).

Decentralisation

The lack of decision on the part of central government over planning formed the background to the two major plans which were made for London's reconstruction, the County of London Plan and the Greater London Plan. These plans were critical, not only for deciding the future pattern of building, transport and industry in and around the capital, but also in providing a lead for authorities elsewhere. The formulation of both plans had as one of their central issues the desire to raise the birth rate and the support of motherhood.

Patrick Abercrombie who was a renowned town planner and keen exponent of the Garden City Movement, produced the County of London Plan with Forshaw, the Chief Architect of the London County Council, in 1943 (Abercrombie & Forshaw 1943). This plan covered what would now be considered to be the inner London boroughs, that is those which came under the jurisdiction of the London County Council. The LCC never formally adopted the County of London Plan, but received it and circulated it for discussion. A newly formed Ministry of Town and Country Planning, which had been set up in advance of legislation in 1943, appointed Abercrombie to make a plan for the Greater London region: this, the Greater London Plan was circulated to officials in 1944 and published in 1945 (SCLRP 1945).

Both the County of London Plan and the Greater London Plan were based on a principle of decentralising population and industry from central London. The existing green belt drawn around London was to be reinforced: the population within that ring was to be reduced and to be persuaded to move to new towns and communities outside the green belt and to 'quasi-satellites' or suburbs within it. The presence of these suburbs was in contradiction with the main aim of both plans, which was to provide planned 'communities' with a balance of employment, housing, shops, social and recreational facilities.

Whilst the County of London Plan gave explicit recognition to the preference of families with young children for houses, it was argued that flats were necessary in the inner boroughs as land costs were high and there was a need for open space. The densities which both plans proposed ensured that local authorities would have to build a proportion of flats in their housing schemes.

The County of London Plan proposed three rings of density. The innermost ring would have a density of 200 persons per acre, the middle 136 p. p. a. and the outermost 100 p. p. a. According to the calculations set out in the plan, each of these necessitated building flats. The 200 p. p. a. ring required that everybody living within it would have to live in a flat; 67% of people in the 136 p. p. a. ring and 45% in the 100 p. p. a. ring would have to live in flats. These densities, although lower than pre-war, followed the same pattern as pre-war London, where the most central areas and some eastern boroughs were most highly populated and the outer suburbs least populated. Abercrombie and Forshaw had expressed a preference for setting the density of the County of London at 100 p. p. a. overall as this would have ensured a ratio of at least 50% houses, but this, they argued, would have meant decentralising too many people and could not be imposed.

Young and Garside, more recent commentators, have suggested that Abercrombie was under pressure from industrialists, particularly those involved in the Port of London, not to decentralise too much industry from London. The war had had the effect of dislocating industry within the Greater London region. Employment declined by 3% between 1939 and 1948 in the Greater London region as a whole. Whilst there were variations within this global figure – the eastern boroughs tended to lose employment whilst the central boroughs gained – the decline was greatest in the manufacturing sector of employment (Young and Garside 1982). Since one of the aims of the County of London Plan was to reduce home-to-work journeys, any reduction in the decentralisation of industry meant a reduction in the decentralisation of population.

Flats versus houses

It was surprising that Abercrombie should have participated in the making of a plan which proposed the building of so many flats. Abercrombie was a keen exponent of the Garden City Movement and indeed had submitted a minority report to the Barlow Commission because he did not think that the main report had gone far enough in advocating low densities.

As was explained in the previous chapter, flats had been built at high densities for workers in Britain from the end of the nineteenth century onwards. This was long enough for some consumer research to have been undertaken: this had indicated that flats were unpopular for bringing up children (Ash 1980). Further surveys carried out during the early years of the war indicated that given a choice, the overwhelming preference was for houses rather than flats (Mass-Observation 1943).

A summary of these surveys was published in the house journal of the Town and Country Planning Association (TCPA 1943). Since the TCPA was intimately connected with the Garden City Movement it was to be expected that this summary should have been published by them, since it supported their views. The surveys varied in scientific rigour and sensibility from academic surveys and those carried out by the organisation Mass-Observation, to the architect Arnold Whittick simply asking his Forces audience after a lecture which they preferred. Of the combined average of 25,000 responses only 5.4% of those surveyed said that they would prefer a flat to a house.

Within this body of evidence there were some suggestive variations, however. For example, the Women's Advisory Housing Council, who surveyed housewives, found that 6% preferred flats. A survey of young women in their teens and twenties, many of whom were single and working in the factories or Forces, found that 14% preferred flats. Furthermore, the Mass-Observation enquiry also found that although women were not averse to flats, a suburban house was the first choice of the majority. However, there did seem to be a consensus that bringing up children in flats was a point of difficulty, especially for women who were charged with the main responsibility for childcare.

Whether or not people would choose to limit their families because they were living in flats is not clear. Evidence from previous centuries suggested that where mortality rates were high, birth rates were high (Gittins 1985): ironically high mortality rates often coincided with overcrowding. Nevertheless, fears about a lowered birth rate pervaded the discussion of both the County of London Plan and the Greater London Plan.

Abercrombie used such fears to support his argument for decentralisation. In the Greater London Plan he criticised those local authorities who wanted to expand industry in their areas, saying that for these people Barlow might never have reported nor had they been touched by the falling birth rate (SCLRP 1945).

The population of the inner London boroughs had in fact declined in the inter-war period. There was an anxiety that if people were encouraged to move from inner London, it would be the young, healthy newlyweds who would go, leaving behind the elderly and the sick. A remark was made in the County of London Plan that it was high time to reverse this general decline and to invest in order to make the inner areas attractive places to bring up children.

The plan suggested that young families might be attracted back to inner London by building a mix of houses, low blocks of flats and high blocks of flats. Old and single people would be accommodated in the high blocks of up to ten storeys. This would leave the lower blocks of flats, of up to three storeys and the houses to young families. There would be open spaces between the flat blocks with communal gardens, allotments, children's playgrounds, tennis courts, flower beds and communal buildings such as nursery schools and social centres.

The eminent American planner Lewis Mumford poured scorn on the efficacy of these proposals. In an article in the *Architectural Review* in 1945 he launched an impassioned critique of Abercrombie's plan. His article was written from an overtly pro-natalist standpoint. Cities, Mumford argued, were in decline because of the higher birth rate in the countryside: 'The great nineteenth century achievement in mass production, the mass production of human beings, has come to an end.' London, Mumford argued, should be planned to reach a target ratio of reproduction of 1.0: '... parenthood itself must become a central interest and duty: the family and the primary group of workfellows and neighbours must become a vital core in every wider association' (Mumford 1945:5).

The densities which Abercrombie and Forshaw proposed were, in his opinion, too high to facilitate bringing up children: Mumford claimed that it would be easier to keep a dog in their proposed housing scheme at Hammersmith than to raise a baby. Houses, he argued, were what Londoners wanted, not flats. With a lyrical flourish he urged that family life be given prime importance in town planning:

> It may be a little while yet before the future Londoner can make
> love to his wife in a summer house in his garden as William Blake,
> certainly no man of affluence did, but that might at least be held up
> as an ideal, if we are as earnestly committed to the rehabilitation of
> family life as we ought to be.
>
> (Mumford op. cit.:8)

If family life and the bearing of babies were to be elevated in status, Mumford thought that this would persuade more women to have children. The temptation for women to pursue work outside the home, which he thought dissuaded women from conceiving, would no longer be as strong:

> Otherwise the first consideration of town planning must be to
> provide an urban environment and an urban mode of life which will
> not be hostile to biological survival: rather to create one in which
> processes of life and growth will be so normal to that life, so visible,
> that by sympathetic magic it will encourage in women of
> child-bearing age the impulse to bear and rear children, as an

essential attribute of their humanness, quite as interesting in all its
possibilities as the most glamorous success in an office or factory.
(Mumford op. cit.:9)

Thus Mumford evoked two stereotypical views of women, the career
woman and the earth-mother, each of which he assumed, were in contradiction with each other. It was the stereotype of all-absorbing motherhood
which Mumford proposed that town planning should support. In order to
achieve his aim of decentralising London into a series of garden cities,
Mumford thought that the nationalisation of land would be necessary.
Despite Abercrombie's pro-natalist views, his plans had failed to con-
vince this arch-supporter of family life.

Household composition

The County of London Plan was based on a wider view of reproduction
than the needs of the nuclear family. Provision was envisaged for those
outside the enclosed circle of male breadwinner, wife and children. This
echoed shifts in social policy as since 1940 children no longer had legal
responsibility for the care of their elderly parents. No longer would sons
and daughters be required to house and financially support their parents;
they could live independently. Nevertheless the provision which was
made in the County of London Plan for childless couples and single
people was in advance of its time.

Abercrombie's justifications for nis proposed densities gave rise to
suspicion that it was pragmatism rather concern for housing all sections
of the population which motivated him. In the Greater London Plan Abercrombie replied to his critics. He argued that to propose decentralising
London below the level set out in the plan was unrealistic. To achieve an
even spread of population through London as had been suggested in a
leading article in *The Times* would have meant there could have been no
provision for schools, open spaces or community buildings. Furthermore,
he suggested that a high proportion of houses within inner London was
unnecessary, since inner London contained a considerable proportion of
one-and two-person households. As no Census had been taken in 1941,
Abercrombie had to base his figures on estimates and he calculated that
approximately 20% of households within inner London consisted of one
or two people. For these households, living in a flat or a hostel 'would be
no social hardship'. Thus Abercrombie was adopting a defensive attitude,
rather than a positive one, in suggesting that flats were a second best
which, if they did not do any good, would not do any positive harm.

Although it was the needs of the archetypal nuclear family, that is man,
wife and two or three children, which occupied the mainstream of planning thought, there was some evidence available which challenged the

dominance of this household type. A wartime social survey had analysed a sample of 1,000 households from different social classes and in different parts of Britain. In this sample only 37.2% of households consisted of two parents and children under the age of 14, which was then the school leaving age: 16.2% of the sample consisted of childless couples. A further 10.3% were households composed of adults without children and 19.4% were households consisting mainly of husband, wife and grown-up children (see Appendix, Table A.1) (Central Office of Information 1949).

The Royal Commission on Population also found considerable variation in household size. The Commission had been formed in 1944 and produced their main report in 1949. Their interest in housing was mainly in the problems of large families: they attempted to make some recommendations for the sizes of houses which they thought local authorities should provide. The Commission used evidence from two surveys, the Wartime Social Survey of 1945 and a Ministry of Labour Family Budget survey of 1937. Using a calculation based on the assumption of a housing standard of one person per room, it was estimated from the first survey that 24% of dwellings should be 5-person, or 3-bedroomed, 29% of dwellings should be larger than that and 48% smaller. On the evidence of the second survey the proportions should have been 27% 3-bedroomed, 33% larger and 39% smaller (Report of the Royal Commission on Population 1949).

This later evidence confirmed Abercrombie's justification: there was a need for a variety of dwelling types and sizes, but to a far greater extent than was envisaged in either of the plans for London. Wartime policy makers were beset with difficulties in calculating future demand, since household structures are based on economic need, availability and marriage patterns as well as individual desire. Indeed, it is probable that in wartime many households were living in arrangements which would not have been acceptable in peacetime, for example, with lodgers or with relatives or friends. Yet it is striking that neither of the two plans for London, nor indeed Mumford's (1945) criticisms included much thought or analysis of future demographic trends. This provides a contrast with other areas of social policy, such as social security, where projections of the future age structure of the population in twenty, thirty and fifty years time formed a subject for discussion.

This lack of projection into the future and of detailed analysis of the present may be partly explained by the exigencies of the post-war housing shortage and the rate of household formation in the inter-war years. Although approximately four million new houses were built between the wars, this rate was only sufficient to meet the number of new households which had formed. Since the average age of marriage in this period was lower than in the nineteenth century and the number of children in each family declined, the rate of household formation rose. Paradoxically towards the end of this period the birth rate declined. Consequently in 1944–5 post-war policy makers would have been faced with a severe housing shortage, a strong pro-natalist sentiment which was held by many shades of political opinion and the memory of the last twenty years of peace in which it must have seemed there was a never-ending supply of young couples eager for housing.

This concentration on the needs of the nuclear family led to a denial of women's and men's needs for housing in a broad sense. Whilst Mumford's complaint was that the plans for London did not allow for a full family life, with the benefit of hindsight and from a different political position we can see that these plans did not allow for a full *human* life since adequate provision was not made for the needs of childless couples, widows and widowers, the unmarried, the separated and divorced and all the other myriad richness of household formation. The legacy of the family policies of the 1940s and 1950s are still present today. They are inscribed literally in bricks and mortar in the form of family flats with no gardens and in a legal system which prescribes that single people under retirement age cannot be deemed homeless and therefore eligible for re-housing. The sum of human misery which has been a by-product, albeit unintentional, of these family policies is considerable.

Mixed development

The ideal of building communities of a mixture of people was, however, uppermost in some politicians' and architects' minds. Bevan, for example, in a now-famous quotation argued that towns and villages should not be divided spatially by class and that the doctor, the grocer, the butcher and the farm labourer should all live together on the same street. He thought that this kind of social mixing was essential so that people could see 'the living tapestry' of a community (Foot 1973:78).

The London County Council (LCC) was pioneering in its promotion of mixed development in their housing estates of the 1950s and 1960s. Both the Chief Architect and the Director of Housing agreed with the idea of creating a community. In a joint report to a government sub-committee they stated:

> In connection with a scheme of development of high flats, it must be borne in mind that the aim is to create homes and that, generally speaking, the old idea of the village or communal settlement is kept well to the forefront.[2]

The way in which the Architect wanted to create the appearance of community was by accommodating different sizes of household in different dwelling forms: old people in ground-floor flats or bungalows, small families in point blocks or balcony-access flats, medium-sized families in three-or four-storey blocks of flats and large families with children in houses or maisonettes with gardens. Thus a balanced mixture of population, in the architect's eyes, would have consisted of various sizes of 'family'.

There were problems in achieving this kind of mixed development on housing estates. The Ministry had laid down as a rule that LCC schemes would not be eligible for subsidy if more than 5% of the accommodation were for old people. The densities in the Greater London Plan meant that a high proportion of flats had to be provided. A further constraint was provided by the housing-subsidy system. The pre-war subsidy system had provided an extra subsidy for authorities to build flats in towns, by relating the subsidy to the higher cost of land. The extra land element was not available for two-storey houses. This meant that the extra cost of providing houses in inner London and indeed in other inner cities, would have had to have been funded from the rates. The LCC was not prepared to pursue this path.

In order to enable the LCC to carry out its policy of building mixed-development housing schemes, the Council's leader, Lord Latham, put pressure on the government before 1944 to include a clause concerning mixed development in its next Housing Bill. The Ministry of Health drew up a clause which would allow cottages to receive the same subsidy as flats on expensive sites, if certain conditions were met. This received approval from the officials at the Ministry of Town and Country Planning, who supported it on the grounds that mixed development would provide the appearance of a balanced community.[3]

The mixed-development clause aroused a great deal of controversy in both the House of Commons and the House of Lords debates on the bills. This was not because members of either House opposed the provision but rather because it raised the whole question of providing extra subsidies for flats. The question of a government bias in favour of flats was raised. Lord Balfour connected this with his fear of socialism:

> That makes me wonder whether there is not a bias here in favour of flats. This was quite common in Socialist circles in Vienna after the last war. They built what are outwardly some fine blocks of flats but inside they are not good flats. They are small, poky little places. The Russians too, about the same time, thought the single family house was a bourgeois conception and they do not like it for that reason. Now it is true, continental thought has gone back to the single dwelling house.[4]

The views that had been proferred in social surveys were stated again – that the overwhelming majority of people preferred houses. Fears for the birth rate were again voiced and Viscount Samuel, in this connection argued that the controversy between houses and flats was an issue of vital importance for the future of Great Britain and for the nation's children:

> ... because this country is now faced, in this controversy between flats and houses, with a problem closely affecting the future generations, and those of us who hold that the principle of family is at the present time of supreme importance and that we should deal with these matters largely with a view to the children of the future and not merely to economic considerations, hold it most strongly that it would be lamentable if the British people were converted into a race of nomadic flat dwellers. What we want above all is to preserve the idea of the English home and have houses which are fit for children to live in, and not only heroes.[5]

Members agreed with the view that a certain number of flats were necessary in large cities. The clause was amended in a minor way, so that houses could be built on expensive sites with a subsidy equivalent to a flat, provided that the Minister of Health agreed.

A certain amount of mixed development was thus provided for in the subsidy system. After 1950 the LCC Architects Department was able to exploit this provision. However, the costs of mixed development were high and the situation was made worse, post-1951, by the Conservative government's squeeze on housing quality. The government manual 'Houses 1955' requested local authorities to raise densities and to lower development costs (MoHLG 1955). This meant that there was more pressure to provide schemes which consisted mainly of flats with few houses.

Had the housing allocation system been organised to represent proportionately each size of household the effects on the raising of children would have been mitigated. In 1951, 49% of the households in inner London contained one or two persons. It seems likely that only a small number of these consisted of an adult and a child aged between two and twelve years, for whom a flat would not be the most desirable dwelling type.

The LCC's allocation system, however, was organised on the basis of need. Need was defined in terms of medical criteria, overcrowding, existing housing conditions and a number of other factors. Each deprivation was given a certain number of points, and extra points were given for families with a scale which correlated number of children against number of rooms. The points system favoured families with children.[6] The LCC, having agreed with the Ministry of Housing to provide only 5% of its dwellings for elderly people, determined the proportions of the rest of its dwellings according to the proportions of size of household on its waiting list. As these were families with children this meant that 80% of new dwellings were for three-and four-person families.

As a consequence some families with children had to be accommodated in flats, which as has been shown, were unsuitable dwelling types for this type of household. Furthermore, other types of household, which could have been accommodated quite happily in flats, were unable to gain access to council housing because of the rationing mediated by the points system. Thus the mis-match between the economics of construction and definitions of housing need worked against women on two counts. Firstly, women in households which were outside of the nuclear family form would have found it hard it have gained access to council accommodation. Secondly women who were in nuclear families and who were charged with the responsibility for child rearing, would have been forced to accept local authority dwellings of an unsuitable nature, that is to say, flats.

Open space

There is another aspect to this controversy between houses and flats. Even with relatively high densities flats are not a necessity, provided that the household size is large enough. Pre-war London had an average residential density of 85 people to the acre and this included low-density suburbs.[7] Part of the reasoning behind Abercrombie's proposal that new developments in London should be mainly in flats lay in his desire to provide public open space and room for other facilities. The Greater London Plan envisaged that there would be 7 acres of playing fields and open space per 1,000 population. This figure would have included school playing fields. The activities listed in the Greater London Plan for these areas of open space were football, cricket, athletics, bowls, tennis, netball and basketball which are predominantly male sports. The figure of 7 acres per 1,000 population was later modified to 4.

Whilst the Greater London Plan proposed sports which are played by both sexes, such as tennis, the impression is of a greater concern for male sports than female sports. Indeed, a later historian, Lord Esher, commented that the planners of the Forties had seen the working man as a Saturday footballer and had not foreseen that the future lay in watching sport rather than playing it! (Esher 1981).

In making such large provision for open space the leisure of working men was taken as being more important than the need of their wives and children for private, useful garden space. This is not to deny the importance of open space: Chapter seven will illustrate the pleasure that being near a park can give. However, it is clear that Sunday football for husbands was of greater concern than usable outside space for mothers with children.

However, it was not only the provision of open space which pushed the densities up in the Greater London Plan. If the entire inner area had been redeveloped at 100 p. p. a. then 200,000 more people would have had to have been decentralised – almost the entire city of Milton Keynes. The planners argued that this would have been unacceptable to industry and business.

The County of London Plan and the Greater London Plan were never formally adopted by the LCC. In 1945 the LCC adopted the planning principles of the County of London Plan in a modified and narrower form. By 1951 the Greater London Plan had been accepted, but again in a modified form, at ministerial level. Abercrombie's plans had laid down the principles for the future development of London.

National legislation

In order to achieve a balance of industry and housing and to achieve decentralisation, the Barlow and Uthwatt Reports had recommended state intervention in land markets. In 1947 the Town and Country Planning Act gave statutory duties to county councils to draw up planning schemes, to compensate land owners and set up a Central Land Board to levy a 100% development tax on planning gains.

This measure was short-lived. After the Conservatives were elected to government, the development tax was abolished, in 1952. In 1954 the provisions for compensation were altered. Thus, from this time local authorities could compulsorily purchase land for public works or they could impose planning constraints on it, permitting only certain types of development. However, they had to pay land owners compensation for either the purchase or for depressing the value. This meant that local authorities had an incentive to buy the cheapest land for their own housing developments, whereas the private sector would develop what the market would take. State intervention in land markets was therefore limited and did not alter the fundamental structure of values caused by a free market situation. In the decade and a half since the Barlow Report the conditions for urban sprawl and congestion had again been created (Self 1957).

Concluding comments

One of the major ideals of wartime social policy had been to preserve and protect the sanctity of family life. Allied with this was a concern, expressed by all shades of political opinion to raise the birth rate. Three Government committees had published reports which suggested that in order to secure the orderly reproduction of the population and raise the birth rate government intervention in land use would have to take place.

These wartime reports proposed controls over the distribution of industry, decentralisation of employment and housing combined with containment of urban sprawl, the encouragement of an efficient agricultural industry and a restructuring of land values which would encourage dispersal and discourage high residential densities. The measures necessary to reach these goals would have meant an unprecedented set of controls over private investment.

In the absence of political guidance on future legislation local authorities had to prepare plans for post-war development. Two important plans were drawn up for London: despite their author being an advocate of garden cities and dispersal, the existing pyramid structure of high land values and residential densities in central London was retained. This, combined with the inclusion of open spaces and social amenities meant that London would be turned from a city of houses into a city of flats.

Critics of these plans thought that the building of flats was not conducive to the promotion of family life and that women would not be persuaded into motherhood by being provided with such a home. There was some empirical evidence to show a widespread dislike of flats for family housing amongst the general population.

Despite the masculine assumption by policy makers and planners that all housing should be built for families and to encourage motherhood, there was a need to provide state housing for other types of household and particularly for women at other stages in their life cycle and in non-familial situations. Childless couples, single women, couples with adult children all required rehousing. These households could have been accommodated in flats: indeed Abercrombie justified his plans on this basis.

The types of state intervention and control which would have ensured post-war development of child-centred garden cities throughout Britain were only partially realised. Whilst this represented a defeat in terms of the provision of family housing it is significant how little attention or importance was placed on the provision of non-family housing in this period. Furthermore such under-provision for the housing needs of those outside the nuclear family has persisted to the present day. This has meant that some families have been condemned to live in unsuitable social housing which easily have met the needs of other types of household, whilst single people under retirement age have hardly been catered for at all.

Having examined the way in which state policies contributed towards a promotion of family life, I shall now turn, in the next chapter, to relations within the family. In the economic sphere policies towards the location of employment and housing provide clear indications of policy makers' intentions and assumptions and these will be considered next.

Chapter four

Women workers and the domestic ideal

In the previous chapter I argued that a wartime ideal of social policy was to support family life and motherhood. In this chapter I wish to look at another aspect of family policy, namely the attitude of the state towards women's employment.

The Barlow Report and Abercrombie's two plans for London both took the relationship of employment to housing as their main theme. Both wished to bring workers nearer to their homes. The focus for all of these documents was male employment. Yet each of them, either by omission or by direct reference, took a position on attitudes to female employment. In their treatment of male and female employment it is possible to see directly the assumptions which policy makers made about the roles which women and men play in the home and to each other.

Housing has a complex relationship to employment in terms of gender differentiation. If it is accepted, and there is considerable evidence that this is so, that the labour market is stratified as between male-employing and female-employing jobs (Barron & Norris 1976; Breugel 1986), then the physical location of housing may either collude with or disturb such stratification. For if capitalism is indeed gender blind and it is through such practices as union organisation, training and the material practice of work that male supremacy is obtained, then the physical availability of workers must play an important part in the process of stratification.

Support of a stratified labour market also serves to deny women access to housing. For if women are confined to impermanent, low-paid jobs then it becomes difficult for them to gain housing independent from men. In this way the housing and labour markets may act to reinforce women into dependency on men.

There were three aspects of the availability of workers to work that occupied the minds of policy makers in the 1940s and 1950s – labour mobility, that is the permanent movement of workers from one part of the country to another: the journey from home to work in any locality: and finally the balance of employment in any district as between male-employing and female-employing industries. These three topics will be examined in detail.

Before considering these arguments the situation of women in the London labour markets will be explained. These provide a background to an understanding of the proposals for employment made in the Greater London Plan and County of London Plan.

Women's waged employment in London 1850–1939

In the transition from craft manufacture to industrial production many processes were (and in some regions still are) put out to homeworkers. In the nineteenth century this was known as 'sweated' or 'slop' work. The London clothing industry in particular ran on these lines, with women forming by far the greater proportion of the labour force. In 1861, in Greater London, the proportion of women to men in the clothing trades was 98,000 : 22,000; in the East End it was 24,300 women to 5,200 men. Even after Jewish immigration in the 1880s, women formed the backbone of these trades (Hall 1962).

There were many types and kinds of women involved in the clothing trade; skilled West End dress-makers, daughters of poor clergymen and others who had suffered a decline in fortune. Concentrated in the East End were those who suffered from old age, illness or were widows or whose husbands were under-employed. One study found that over half the clothing homeworkers in West Ham in 1907 were the wives of builders, general labourers and dock labourers, all of whom were employed on a casual basis (Hall 1962).

London, of all employment centres in Britain, had the largest pool of unskilled labour, which despite being unskilled, could nevertheless be trained to operate a simple machine. This pool, which was in existence since 1850, was made up primarily of women and secondly of immigrants, who, of course, would also include women. Employers in service industries would be attracted to areas of heavy male employment such as the docks.

One further example of this type of low-paid home work, greatly exploited, comes from housing history. White describes how, in the pre-First World War tenement block he studied, one woman had made a living by making cigarettes at home:

> Jenny Schlom's mother made cigarettes at home and rolled the paper cigarettes for 4d – *four pence* per 1,000. At financially difficult times she got work from her brother, a cigarette maker, and Jenny would take the finished cases and cigarettes to her uncle at Bow.
>
> (White 1981:238, emphasis in original)

From the end of the First World War there was a permanent acceleration of the trend towards the employment of women (Wilson 1977). In the
inter-war period more women entered the waged labour force. New employment opportunities opened up: shop assistance and clerical work and
for the educated, teaching and nursing. London, as a commercial centre
and a centre of population could support light industry. These, together
with service industries, provided a variety of employment for women.

Durant found, in her study of an LCC council estate on the periphery
of London, that newly installed factories nearby provided jobs for women
and young people. These were not 'career' jobs. As many as 60% of
women on the estate were registered at the local labour exchange (Durant
1939). Hall (1962) argues that as in the nineteenth century, new industries
were attracted to London by its market and by the local labour pool, in
particular the pool of female labour. He cites the chairman of Slough
Estates Limited who described the type of worker employed in new factories as 'semi-skilled. Machine minder is the appropriate term ...
commonly this was female labour' (p.166). Hall concluded that the difference between pre-and post-First World War industry in London was
that the range of industries was much greater.

The LCC's evidence to the Barlow Commission stated that a greater
proportion of women than men were in employment in London than elsewhere in England and Wales, compared to the total male and female
populations. Whereas 20% of all males employed in England and Wales
were working in Greater London, 25% of all females employed in England and Wales were working in Greater London. The proportion of
women in employment in Greater London had increased from 32.3% in
1921 to 36.1% in 1931 whilst, during that period, the total female population of Greater London had declined.[1] The LCC concluded that only a
small proportion of this increase was due to the growth of manufacturing
industry in London:[2] the other industries which had supplied increased
employment for women were personal services (which included hotels
and catering), distributive trades, clothing manufacture, provisions, electrical installations, cables and apparatus, entertainment and sport.

The Second World War changed the nature of everyday life. As such,
for the duration, women's employment was also altered. The next section
will examine the nature of these changes which mirrored those that occurred during the First World War.

Women's employment 1940–1947

In 1941 women were called upon to contribute to the war effort, for the
first time ever on a compulsory basis. The Registration for Employment
Order and the National Service Act (No. 2) meant that young single

women had either to stay in the protected industries, or be conscripted, or move to do essential work elsewhere in the country.

Regulations about women's employment were drafted in such a way as to leave undisturbed existing notions of women's domesticity. Married women, for example, were regarded as 'immobile'; that is, they could not compulsorily be relocated, even if they had no children and their husbands were away in the Forces. Men were assumed to require someone to look after them in their houses. A judgement which was made by the Ministry of Labour with reference to cases of young, single women looking after widowed or separated fathers stated:

> There may be exceptional cases in which a man is able to live alone in his house and to perform all the duties connected with the house but they are rare. As a general rule the presence of a woman is essential to the household.
>
> (Cited in Allen 1983:406)

This judgement made explicit women's supposed duties and men's assumed incapabilities!

Women joined the waged labour force in unprecedented numbers during the war, to do both part-time and full-time work. Two and a half times more women were mobilised for industrial employment than in peacetime (Land 1976). As in the First World War women were more visible, since some of them were entering men's jobs. Traditional male industries such as engineering and transport were 'diluted' with female labour. Some of these women withdrew at the end of the war: for example the National Federation of Women Workers agreed to withdraw its members from jobs claimed by the Association of Skilled Engineers in return for joint wage negotiations (Wilson 1977). Similarly, these jobs did not give them equal pay and were regarded as temporary.

More effort was made to accommodate the needs to replace women's work in the home than in the First World War. Consideration was given to childcare, food and shopping. Such measures that were taken were, however, limited. A policy of providing nurseries was only pursued until 1941, when, because of conflict between the Ministry of Labour, who were in favour of nurseries and the Ministry of Health, who were not, it took second place to the provision of private childminders (Summerfield 1984). By 1944 three-quarters of the children of women war workers were being looked after by private childminders.

A school meals service was set up, factory canteens provided and British Restaurants – that is local-authority-run restaurants were inaugurated. This experiment, although popular, was restricted in scope and started late in the war (Roberts 1984). The Ministry of Food was reluctant to place restrictions on retailers to help women cope with long factory hours and the problems of rationing and long queues. Employers were encouraged to provide leave of absence for shopping, a scheme which dove-tailed with the provision of part-time work.

Women war workers, far from heroically managing the home front despite enormous difficulties as portrayed by wartime publicists, took time off work. Rates of absenteeism amongst women workers and in particular married women workers were high. Where necessary, when the burdens of the 'double shift' became too great, women workers simply left waged employment (Summerfield, 1984).

In the immediate post-war period, from 1945 to 1947, many jobs which women had been doing disappeared since the work had been created by the war itself. The Nuffield College Reconstruction Survey of 1941 had predicted that between half and three-quarters of a million women would want to leave their jobs at the end of the war.[3] By the end of 1946 there were still 875,000 more women in the working population than there had been in 1939 (Land 1976). Thus while the wartime peak of women's employment had been temporary, as in the inter-war years the trend was upwards. Although women had entered what were considered to be male jobs during the war, the sex segregation of the labour force was left unchanged. Apart from teaching and the Civil Service, where the marriage bar had been lifted, women were mainly confined to low-paid jobs which were defined as unskilled.

Labour mobility

The Barlow Report had considered the uneven location of employment and population throughout Britain. It had recommended that employers be encouraged to take their industries to depressed areas and to be prohibited from setting up new industries in London.

The Hopkins Committee on Re-Construction Priorities, which was set up in 1943 by the wartime Cabinet also considered the problems of industrial location and labour mobility. The problems of labour mobility were considered in relation to each of the sexes. The administration of wartime controls over industry had provided some experience of the problems of relocating workers. The Committee remarked that they had found married women tended to stay at home or move with their husbands. Single women were prepared to move because it was wartime but apparently their domestic responsibilities and wish for home had been great.[4]

The Hopkins Committee concluded that the barriers to mobility for workers were having a skill not in demand elsewhere, local ties, optimism that the situation would improve and an ability to live on the dole! The committee thought that inducements had to be offered to overcome these obstacles. These inducements were firstly, a definite job with good wages and prospects and secondly, arrangements such that workers could be assured that their wives and families could come with them and be provided with affordable housing. Women's domestic responsibilities and ties were not seen as barriers which could be altered or changed: instead a link was made between housing and employment which was that men had to be supplied with their home comforts if they were to be persuaded to move.

The importance of housing was stressed by the Hopkins Committee. They emphasised the necessity for a policy to be declared early after the war ended of providing a large number of houses to let at a low rent if labour mobility were to be encouraged. Furthermore, they argued, cheap housing was not enough; a good environment should also be provided by a proper system of town and country planning.[5]

The Scott Report had also been concerned about the effect that poor rural housing conditions were having both on the birth rate and on the take-up of agricultural employment. In the Committee's view labour mobility was tied up with housing conditions: the assumption was that whilst the man's job was important in terms of providing a family wage, the woman's job – which was taken to be domestic work – was also important in terms of the stability of the household. For a workforce to settle, the Committee recommended, both the conditions of work and of agricultural housing should be improved (Scott Report 1942).

The idea that a male workforce would only be persuaded to move if a guarantee of its home comforts were met by the provision of housing and essential services was carried into the Distribution of Industry Act 1945. The assumption here was that women were domestic workers in the home and their needs for good conditions of employment would be met by a supply of high-quality rented housing. There is some evidence of a discussion in the Civil Service during the drafting of the Bill as to whether there should be an explicit reference to the needs of workers' dependants in an enabling clause which gave the Board of Trade powers to provide housing and essential services for relocated firms. At final draft, the clause was left to refer vaguely 'to the needs of the area' and housing and services were only to be provided for key workers.[6] Industrialists had to apply to the Board of Trade for an Industrial Development Certificate before they could establish a new plant: these controls were later included in the Town and Country Planning Act 1947.

The County of London Plan and the Greater London Plan were prepared in advance of these decentralising activities by central government. Abercrombie's proposals were, however, broadly in line with their recommendations. The situation of London was unique, for, unlike the rest of the country, the local authority was faced with the problem of moving work, homes and workers; furthermore a significant percentage of these workers, as has been discussed, were female.

The Greater London Plan (SCLRP 1945) recognised that members of the household other than the man might be engaged in waged employment. A recommendation was made that both industry and business should relocate to new 'balanced' communities to enable more than one household member to work. Thus the varied needs of the family might be met. Yet there was no doubt that the man's job, as chief wage earner, was regarded as the most important in the household. This assumption was hinted at in a comment that good transport and a variety of employment should be provided for situations where the chief wage earner changed the location of *his* job but did not want to change *his* home.

That male jobs were valued more highly than female jobs was made explicit in a discussion of which industries should be allowed to relocate first. For example, it was argued that removing a printing works or an engineering factory would automatically cause a migration of labour. These enterprises, it was thought, would have trained their staff and would wish to take them with them. As these workers would be 'chief householders', their families would automatically follow them. Conversely, if a biscuit or a clothing factory were to move, it was argued that this would not induce a migration of labour, for these industries employed only 'unskilled' labour, which the authors thought would only be young girls who would not wish to change areas.

This last comment also carried the assumption that it was not married women with children in the household who worked, but young single women. The Greater London Plan clarified this point in a discussion of the conditions necessary to induce migrant labour to settle in their new surroundings. Married women were not considered as industrial workers, but as wives. They noted that the main reason for workers who had migrated moving back to London was that their wives did not like their new surroundings. This in their view emphasised the need to provide good cheap shops and other social and recreational facilities. As wives it was assumed that women's main task would be to service their husbands: that women could be the sole or only wage earner was not considered, nor that the wives' wages might be crucial to the household's survival.

The figures for the number of households which were either wholly or partially dependent on a woman's earnings are not readily available for the 1940s and 1950s. However, figures for the 1970s give some idea of the scale of women's responsibilities. In 1971 one in six of all households were dependent upon a woman's income (Land 1980). In 1974, 7 million working wives contributed on average 25% of the family income and half a million were the sole or primary earner (Land 1979). The Family Income Supplement was introduced in the 1970s, a benefit which designed to raise the living standards of a family household up to a minimum level where the full-time wages of the head of household were so low that not even a minimal standard of living could be achieved. The Department of

Health and Social Security carried out some research before this benefit was introduced which showed that approximately 200,000 households would have been eligible for this benefit if the wives of male heads of households did not also go out to work (Barrett & McIntosh 1980).

Journeys from home to work

The journey from home to work provided a vital part of plans for decentralisation. London's transport problems had long been a cause for concern. Young and Garside have noted that discussion of London's problems in the 1930s made comparisons between the city and a human organism, supposing that it would become so congested that eventually it would rupture and collapse unless the pressure were reduced (Young & Garside 1982). This imagery of enfeeblement was taken up in the Greater London Plan where it was suggested that London's contemporary transport problems had made Londoners 'a race of straphangers'.

The Barlow Commission had emphasised the need to reduce home-to-work journey times for economic and social reasons. The Greater London Plan, in its proposals for decentralisation, took this as an aim, but recognised the problems of individual household members working in different places. A statement in the Plan conceded that as long as there was a free market in jobs, and that family members had different wishes, it would be impossible to prevent a large number of cross-town journeys.

The Greater London Plan supported masculinity by assuring greater priority to male jobs in plans for decentralisation. Male home-to-work journeys would, if the plan were implemented, be considerably diminished. Meanwhile, parts of inner London were to be reconstructed on the lines of 'neighbourhood units'. The County of London Plan described in detail how such units would be planned. A local industrial zone would be provided for each area and would service a number of neighbourhood units. In this zone there would be laundries, bakeries and other service industries plus local industries. Although a small number of workshops would remain, their numbers would be carefully controlled. The intention of the plan was to reduce the numbers of backstreet workshops and to remove these employers to flatted factories, 'several stories' high on the industrial estates.

Since it would have been likely that this type of 'small concern' or workshop would have tended to employ more women than men as, for example, in the East End clothing trade, this re-grouping would have had the adverse effect of increasing women's journey from home to work. Judging from the diagram provided, the distance involved would have been in the region of three to five miles. Shops in the neighbourhood unit were also to be grouped together. Local shopping centres would have been provided, each retailer chosen to avoid 'over provision' of shops with 'excessive competition' between them. Larger shopping centres would remain. Not only are women employed by shops, but shopping takes up a major part of running a household. Thus within reconstructed areas, major sources of female employment were to be relocated in a way which would increase the journey from home to work for women – in direct contradiction to the aims of decentralisation.

Balance of female and male employment

A consideration of the different types of industry to be provided in any one area leads to my third theme, that of the 'balance' of industrial employment between men and women. I have suggested that London was a unique location in terms of its pool of unskilled labour, made up of women and male and female immigrants. This labour force, it has been suggested, pre-existed the boom of the inter-war years, the difference between pre-1918 and post-1918 being in the type and range of industries which this labour force and market could attract. The Barlow Commission also noted that industrial expansion following the First World War had provided employment for a relatively high proportion of women, although they attributed this to a change in employment rather than a reinforcement of an underlying trend.

The fact that the Barlow Commission did see the expansion of female-employing industries as new, and not as a continuation of a previous pattern, is significant. In the Introduction, I referred to fears expressed by Engels that the employment of women and children in Lancashire cotton mills, to the possible exclusion of men, would mean that the family would be 'turned upside down'. This did not happen – in fact men gained ascendancy within the mills and women were later confined to low-paid occupations. The ideal of the family wage took over.

What was seen as an increase in female employment in London in the inter-war years raised old fears. The Scott Report, for example, weighed up the pros and cons of attracting light industry to rural areas. On the one hand, these industries could provide employment for women and young people: this would have the advantage of increasing the family income and reducing the isolation which many rural women suffered. On the other hand, male workers could feel discontented if their wives and sons and daughters earned more than them and in better conditions. Whilst the Committee considered that these difficulties could be overcome by improving the wages and conditions of agricultural labourers, further perils were still present, not least in removing married women from their home and presumably providing domestic comforts:

> The provision of alternative occupations for women and young
> people would be of advantage but certain dangers might be

involved, as for instance, the attraction away from agriculture of the younger generation and the causing of discontent, at least among their menfolk, by attracting women from their households for full-time paid employment.

(Scott Report 1942:69)

Accordingly, the report recommended that new light industry should be located in existing or new towns in country areas with careful controls over its siting.

Abercrombie also warned of problems which might ensue if trading estates were too greatly used as a means of industrial development. Trading estates, he reported, tended to attract firms which were unstable:

...frequently these concerns are large employers of young persons and low-paid female labour. These conditions can create very difficult social problems, which are not conducive to a sense of citizenship, unless the number of such firms in any one area is limited. A balance should be built up with stable concerns employing skilled male labour.

(SCLRP 1945:51)

Although Abercrombie used the gender and class-neutral rhetoric of citizenship to make his point, there is an echo of a fear of the 'home turned upside down', of the existing order of familial relations dissolving.

These examples provide evidence for the reasoning behind the expressed need to provide a 'balanced' distribution of employment between male-employing and female-employing industries in any one area. The Barlow Report had stressed this need, arguing that also such a diversity of industry would lessen the risk of unemployment 'for the worker and his family...'. Large towns were felt to be the best kind of settlement from this point of view, in providing the right kind of proportions of different industries.

The word balance implies an equal relationship, in the words of the *Oxford English Dictionary*, an 'even distribution of weight and matter'. In this instance, it seems that the phrase 'balance of employment' is not used to denote an equality between women and men, but rather a perpetuation of existing inequalities. The domination of men over women in the labour market was to be supported in order to avoid upsetting the existing relations of male domination and female subordination within the family.

The argument that providing a suitable distribution of male-employing and female-employing industries would lessen the risk of unemployment was also used in the command paper *Employment Policy*, which was published in 1944. This document set out a policy for maintaining a high level of employment, drawing on Keynsian economic theory, which proposed government intervention in promoting increased consumption during a slump (H. M. Government 1944). As such, this document laid a consensus on economic policy which was to last for the next thirty years. The paper discussed the need to avoid local patches of unemployment. The method recommended for doing this was to encourage a balanced distribution of industry. An industrially balanced area was defined as one which neither was over-dependent on a single industry or group of industries, nor where the industries were concerned solely with the export trade, nor where demand fluctuated unpredictably nor, finally and in this instance crucially, where there was employment mainly for men or for women.

Thus a shift had taken place in policies towards a family-wage economy. Whereas the consensus which had dominated capital–labour relations from the mid-nineteenth century onwards was that of an ideal family consisting of male breadwinner and economically dependent wife and children now, in 1944, women's employment was recognised to be important in increasing a family's income. A slide into poverty, should the man become unemployed, could thus be overcome. This shift did not endorse role reversal: rather it represented an acknowledgement of existing conditions in a stratified labour market in which men were dominant.

Government policy moved still further in its recognition of the importance of women's employment three years later. The Economic Survey of 1947 noted that the labour force of approximately 18 million men and women fell far short of what was required. It therefore called upon women to take up industrial employment to aid 'the national effort'. It also exhorted employers to help women with domestic responsibilities, i.e. married women by offering part-time work, as in wartime. This call for women to enter industry was part of a national drive to increase production for export (H. M. Government 1947).

Despite this overall plea for women workers, the Board of Trade still kept a watchful eye over the distribution of male-employing and female-employing industry. Its Policy Study Group found that it was firms which employed 'a considerable proportion of females' which were most likely to move to the development areas of the north-west. This finding gave rise to some concern about whether areas outside the development areas were attracting enough male-employing industry.

Central government intentions to provide sufficient jobs for women but not to upset existing relations between the sexes within the labour market and the family were reflected at local level in the County of London Plan and Greater London Plan. Abercrombie and Forshaw made these intentions explicit in the County of London Plan. In a section discussing the planning of new areas they remarked:

In the development of industrial estates close regard must be had to the type of industry contemplated. Light industry is a big employer of women and young persons; hence some of the heavier variety is desirable if industrial stability is to be achieved and work provided for the young and adult male population.

(Abercrombie and Forshaw 1943:97)

For this reason they argued that new industrial estates should be located on the fringe of dormitory suburbs where employers could draw on a pool of female labour. Abercrombie and Forshaw argued that new industrial estates were not suitable for new towns, because of the female labour force they would attract and were only acceptable if they contained a reasonable amount of male-employing businesses.

In existing areas Abercrombie and Forshaw thought that whilst some industry needed to be decentralised, this should not be taken so far as to upset the balance of labour: 'Care must be taken not to upset areas where employment is well balanced as between men and women' (ibid: 96). Accordingly they suggested that some industries were not to be decentralised. For example, in the East Thames, where the main industries of docking and ship-repairing were located, electrical work was to be kept since it was a great employer of women.

The types of industries which the authors of the plan thought should be moved outside London were predominantly male. They were engineering (male dominated), food factories (female), furniture (male) and printing (male). As has been argued previously the theory behind this set of priorities was that men were more likely to move with their families than women. In the Greater London Plan, Abercrombie argued that if male-employing industries were decentralised to new towns first, then female-employing industries would be attracted at a later date. As a consequence he suggested that jam and biscuit manufacturing companies in Bermondsey should be moved to the suburbs or outside London 'near an ample labour supply' (Abercrombie & Forshaw 1943).

So far central and local government policy intentions have been considered. In the next section a brief look will be taken at how far these goals were achieved.

'Balanced employment in practice'

Contrary to the aims of the County of London Plan, the LCC had acquired sites early on in the Second World War in the Green Belt. Given the housing shortage, the Council felt that it would be impossible not to develop them. Since one of the largest suburban estates which they had built before the war, Becontree, had failed to attract tenants because of lack of local employment, the provision of jobs on these 'quasi-satellites', as these settlements became known, was of keen concern to the Council.

Difficulties were experienced. First of all lack of labour and materials after the war meant that priority had to be given to building houses rather than factories.[7] Secondly, delays occurred in the granting of industrial development certificates.[8] In the face of these difficulties firms were initially slow to offer to move to these new sites.

By June 1950 the council officers were able to report some progress on the relocation of firms to the quasi-satellites. The majority of such firms seem to have been male employing. For example at Debden there were firms making scientific instruments, drawing-office materials, steel fabrication, leather travel goods and precision engineers, piano makers and book-binders. At Hainault there were furniture makers, packing-case makers, engineering and joinery firms and coach makers. There were two businesses which could have employed women – a confectionery manufacturer and another described as 'sound and television'.[9]

The impression that the industries which decentralised employed mainly men is born out by Jefferys' study of South Oxhey, another LCC quasi-satellite. The study, which was primarily about health, included a section on working women and their employment problems (Jefferys 1964).

Only a relatively small proportion of the married women, one in five, had been working before they moved onto the estate. This low proportion may be explained by the fact that nearly 40% had at least one child under school age and 80% a child aged 14 or less. Nearly all of them had given up their jobs in London when they moved.

The estate was completed between 1948 and 1952. By 1954–5, 40% of the married women at South Oxhey were in employment. Just under half were in full-time jobs, the remainder worked for 30 hours a week or less. For most, their employment was not near to their homes. Jefferys found that only one in four of the wives had found work in the estate itself; the majority (60%) were working within four miles of the estate and 10% had to go farther away. The full-timers, on the whole did more skilled work, whereas the part-timers were employed mainly as cleaners, shop assistants and in canteens and laundries.

Jefferys concluded that the lack of employment for women had the effect that some of the women who would have liked to have taken up waged employment could not because no work was available locally. Further, other women had to take jobs at a longer distance away from home than they would have liked, leaving them with problems of extra travelling expenses plus difficulties with housework and childcare.

Thus, the emphasis on the provision of jobs for males had the result that women were more limited in terms of places of employment. Further-

75

more, as a pool of labour, and this is speculation, they potentially became fair game for any manufacturer who wished to move his plant nearby.

Whilst sources of work for women were not being provided on the out-county estates, ironically, employment for women increased dramatically in inner London. The main thrust of employment policy had been towards reducing manufacturing industry in London: whereas in the period 1945–1957 office employment grew without any restriction. Hall (1962) estimates that in the period 1945–1951 office employment in London grew by between 29,000 and 68,000 jobs. This increase was unforeseen: until 1951 no provision was made for the growth of office development. In 1951, 60% of all office employment was for women and by this time office work provided the largest employment of any female occupation.

Concluding comments

Employment policy had moved in its assumptions and predictions from pre-war practice. Women's employment was to be encouraged, even married women's employment, to boost the export drive. The form that this employment was to take was not, however, to upset notions of appropriate gender roles; the home was not to be turned upside down by any hint of role reversal. The basis of post-war employment policy was that men 'naturally' held dominant positions in the labour market – male jobs were skilled and relatively well-paid, female jobs were unskilled and low-paid. The goal of policy was to ensure that an adequate mix of employment was provided in each geographical locality to support and reinforce this sexually stratified labour market. As in the post-1945 social security system, documented by Land (1976), there was a lack of recognition that women might be the sole economic supporters of a household, in particular in households where an adult male might be present.

Since work and workers were not evenly or logically distributed throughout the country and in cities, post-war policy had to grapple with the problems of labour mobility. In the main, as Self (1957) commented, work was brought to the workers. In the case of London both work and workers had to be moved. The provision of good-quality housing to rent, with adequate services and recreational facilities was thought to be vital to this process. Only by raising the standard of women's domestic conditions and ensuring the breadwinner a contented home, could key workers be induced to move. The basis for this policy was an assumption that women's domestic responsibilities were an unquestioned facet of social life.

Thus, despite an accommodation for married women in post-war planning, the dominant assumption that women were primarily associated with the domestic sphere persisted. In terms of labour mobility this meant that women would only move home geographically in order to follow their husbands and not on their own account. However, a beneficial corollary of this assumption was that the importance of the provision of cheap rented housing, shopping and entertainment facilities in connection with work places was recognised.

A concern was expressed in reconstruction planning to reduce journey times from home-to-work. It was male home-to-work journeys which provided the focus of attention. I have suggested that, ironically, ideas about concentrating work places and shops in neighbourhood schemes meant that journeys from home to work and journeys connected with domestic work were, if anything, lengthened for women.

A fear of role reversal underlay discussion on the balance of employment in particular areas. Accordingly, an emphasis was placed on the necessity of providing well-paid skilled jobs for men. Female employment was considered in terms of providing unskilled, low-paid work for young unmarried women and older married women whose children had grown up.

These policies towards physical planning were in contradiction to post-1947 employment policy which sought to draw married women into waged work. Jefferys' study of South Oxhey recorded, in passing, the need which married women with pre-school or school-age children experienced to enter employment.

This discussion of reconstruction planning, then, illustrates that a powerful set of assumptions about gender divisions were implicit in physical provision for employment. On the one hand, these ideas were part of a set of dominant values which provided a framework for women's lives. On the other, these values and ideas were, in some cases, in contradiction to employment policy post-1947 and to the needs which some women themselves experienced to combine waged work, motherhood and housework. Almost unwittingly, it seems, employment opportunities for women did present themselves through the growth of office building in central London in the 1960s. This growth in office building had neither been anticipated nor planned for and was contrary to more general plans for decentralisation.

Chapter five

A woman's home is her factory

The experience of the Second World War had imposed some levelling of class difference upon British society. The unexpected election of a Labour government also raised the possibility of a more egalitarian exist-ence. At the end of the war, there was a general desire for a more equal society than had existed during the 1930s (Marwick 1982). There was both a levelling down, in a middle class now unable to employ domestic servants and a levelling up, with better housing standards and more edu-cational opportunities for the working class.

This chapter will examine this levelling process with reference to the raising of working-class housing standards. Whilst housing standards may seem to be a political issue connected with class, they also are inti-mately connected to perceptions of gender difference: in this period to notions of the housewife's role in the family.

The Second World War had also forced a degree of communality upon the population. Men and women were called up to live an institutional life in the Forces: evacuated children were billeted with strangers: people were forced through lack of housing and through economic necessity to either become lodgers or to take in lodgers: a small number of services such as soup kitchens, canteens and nurseries were set up. Although such communal facilities introduced those who used them to new ways of or-ganising domestic life, they did not necessarily wish them to continue. There was evidence of some men, at least, who wished to return to pre-war patterns of normality (Roberts 1984). The LCC made two tentative attempts at providing collective facilities, in the form of playgroups and and laundries. Reactions to these communal services had a class dimen-sion which will also be explored.

Post-war housing policy in terms of standards was a subject of discus-sion during as well as after the war. Housing standards were seen not only in terms of bricks and mortar, but also in relation to domestic morality and to standards of housewifery. This aspect of standards also had a class dimension in which middle-class aspirations about women played a strong part.

Post-war egalitarianism

The election result of 1945 was a surprise, not least to the Labour Party itself. Churchill, the powerful wartime leader, had been expected to bring his party to victory (Addison 1982). Many reasons have been suggested for this unexpected turn of events – the influence of intellectuals before the war in writing and broadcasting about democracy, equality and justice; the effect of the war effort with its emphasis on fair shares for all; the influence of the Forces Education Programme which was compulsory for all ordinary soldiers and was devised and taught by liberals; and finally the effect of the war in terms of providing a disruption and holding out the possibility of a better world (Davies 1984; Calder 1965).

The way forward for an increased measure of economic and social equality had been laid during the war with the publication of the Beveridge Report. Beveridge's recommendations – which he saw as part of a campaign to rid Britain of the five evils of 'Squalor, Ignorance, Disease, Want and Idleness' – ensured at least a minimum standard of living above the poverty line. However, his proposals for social security did not propose an equality between women and men. Land has shown how Beveridge's recommendations carried a number of powerful assumptions about women's dependence on men and on their inequalities in the labour market (Land 1976, 1979).

Allatt has shown that a similar presumption of inequality also pervaded the Forces educational literature. In order to achieve social cohesion amongst the ranks who might come from different social classes, instructors addressed the men as men and appealed to a common, masculine solidarity. Women, and particularly women working, were seen as a 'problem'. The family was posed as a value in itself, worth fighting for. Women were identified as housewives and mothers, although the prospect of married women undertaking some part-time work post-war was also admitted. Feminist ideas of equality were closely scrutinised and challenged. Allatt notes that the construction of new houses was often the sole plea for change in women's lives post-war (Allatt 1983).

Moving into a new house did not mean a simple change in physical location. One Forces booklet suggested that certain people had to be socialised to move into new housing estates (Halton 1943). The idea that the acquisition of new values was part and parcel of a move up the housing ladder seems to have been prevalent before the Second World War as well as after it. Before the war, slum dwellers were stigmatized because they lived in poor accommodation, and were thought to be slovenly and dirty in their habits and loose in their morals. The Chief Constable of Glasgow wrote approvingly of some tenants who had been rehoused on the periphery of the city in 1925:

> When they first came, the women folk showed signs of slovenliness and untidiness. Frequently they went shopping bareheaded, hair disarranged, faces unwashed, shoes unfastened, and sometimes without boots or shoes. A shawl thrown carelessly over their shoulders afforded them a small measure of protection from cold and rain. After a very short time, however, these women on seeing better dressed members of their sex moving about in the vicinity, gradually began to take an interest in their personal appearance. The consequence now is that these quondam slum-dwellers are now comparatively clean and smart as they set out to do their shopping. The wearing of hats is a very noticeable feature in the new garb of these exiles from slumdom.
>
> (cited in Mackintosh 1952:209)

Mackintosh, a public health inspector writing in the 1950s, also noted a change in the attitudes of people rehoused onto new estates. He interpreted the putting up of new curtains as a sign of an increased self-respect 'at once barriers and banners'. He also thought that women entered into the life of the community more, taking an interest in their children's clothes, in parent–teacher relationships and in the health services.

It was not only women who were affected by rehousing. Men, also changed, but in a different way. Mackintosh cited a survey he undertook in the 1920s. There he found that the men, husbands, spent more time at home, gardening or doing odd jobs about the house, or went out to play bowls rather than spending evenings in the pub (Mackintosh 1952).

The change in habits which was to be promoted by rehousing was to be a transformation. Rehoused slum dwellers acquired the values of thrift, sobriety and self-respect. These values had been associated with the upper echelons of the working class as well as the middle class (Cockburn 1983). Social status for men relied on their masculinity being supported by women who conformed to a particular model of femininity. An ideal was proposed of the respectable but responsible woman, whose main role and interest in life would be caring for and servicing her family, at home and in the local area. Such a woman would be bound by the mores of the 'respectable' working class – sexually chaste, modest, clean, neat, tidy, quiet, private – and by the virtues of the liberal middle class – responsible, active and involved in the 'community'. This ideal, however had its costs, economic as well as social.

British housing in 1945

Despite the boom in owner occupation which had taken place in the 1920s and 1930s Britain still faced an acute housing shortage after the Second World War – 200,000 houses had been completely destroyed and 3.5 million damaged. No slum clearance had taken place since 1939 and the rate of new house building decreased sharply. The British population had increased by one million during the war, even allowing for the numbers of deaths associated with it. Furthermore, there had been a rise in the number of marriages in 1939 and 1940 and an increase over the inter-war years in the years following in the number of marriages per year. This meant that there was a high rate of household formation and an increased demand for housing (Merrett 1979).

The condition of much of Britain's housing stock was also poor. No census was taken in 1941 but the 1951 census revealed the poor state of many houses: 38% of households had no bath, or even use of a bath at all; 8% of households did not have an outside or inside toilet (Murie 1983). A survey in 1942 revealed that 30% of households earning less than £300 a year (a majority of the population) had no other means of heating water than in pans on the stove (Forty 1975). Given this level of deprivation in essential amenities, levels of disrepair were secondary.

Although large numbers of houses lacked essential services such as piped hot water, gas and electricity, the mass production of domestic appliances had developed in the inter-war years, as was discussed in the first chapter. This meant that the disparity in the amount of labour involved in looking after a house which had the benefit of electricity and therefore running hot water, vacuum cleaners, an electric or gas stove and a built-in bath was significant in comparison to a house which had only coal fires, gas lighting and no means of heating hot water other than pans on the stove. It was this kind of difference which in its turn provided a material base for class divisions which formed the background to central government proposals for raising housing standards.

The Dudley Report

During the Second World War, as in the First World War, central government set up a committee to set new housing standards for post-war developments. At its first meeting the committee broadened its terms of reference from reviewing the plans in the existing manuals for council house architects to reviewing virtually every aspect of standards for houses throughout the country.

The Dudley Committee started life with eleven male members, two female members and a woman architect joint-secretary. Some women's organisations complained to the Minister about the male bias of the committee. A suggestion was made that, as with the Tudor Walters Committee a Women's Sub-committee should be set up. Opposition to this was met from some women's organisations; which were not recorded, but a possible reason for their dissent was that the Women's Sub-committee to the Tudor Walters Committee had been treated in a derisory manner by the main committee. Four further women members were co-opted on to the Dudley Committee making a total of six women to eleven men. The committee first met in 1942 and reported in 1944. In the course of its work it consulted fifty-seven organisations, sixteen of which were women's groups. These ranged across the political spectrum from the Union of Catholic Mothers to the Standing Joint Committee of Working Women's Organisations. The women's groups which were consulted took their role seriously. Three groups carried out surveys using questionnaires. The Women's Housing Advisory Council sent out 40,000 questionnaires, the Society of Women's House Property Managers collected evidence from a questionnaire put to tenants on their estates and the Women's Group on Public Welfare also carried out a survey. The evidence from the Women's Housing Advisory Council was taken most seriously by the Dudley Committee, who arranged a special meeting to hear their evidence.

As after the First World War, following the publication of the committee's report, a design manual was produced by central government based on it (CHAC 1944). The *Housing Manual* (MoH 1944) commented that much of the evidence brought to the Dudley Committee was from the 'housewife' or 'consumer's' point of view. This certainly seems to have been the case in that it was those recommendations which women made as housewives which the committee paid most attention to rather than representations made on behalf of single women's organisations. In Chapter three the contribution towards an emphasis of post-war housing policy on the family – almost in effect a family policy – has been described.

What is of interest here is the type of house which the Dudley Committee sought to promote and how they saw the role of the 'housewife' within it. Ravetz (1984) has argued that the 1950s saw a submergence of previous divisions between middle-class women and working-class women and the emergence of the classless 'housewife'. Certainly some attention had been given to the lot of the 'tired housewife' during the war and women architects such as Jane Drew vowed to take the drudgery out of housework (Drew 1944).

The attempt to reduce the labour in housework meant that attention was focused on the kitchen. As Wilson records, the theme that the housewife's workshop is her kitchen was popular, unfavourable comparisons being made between the conditions of industry and poor housing (Wilson 1980). The Standing Joint Committee of Labour Women's Organisations took up this theme in their evidence to the Dudley Committee: 'The "tooling" of the kitchen is as worthy of careful attention as is given to the design and arrangements of machines in a modern factory.'[1]

The purpose of careful design and consideration for the appropriate relationship of kitchen to dining to living room was to save labour and frustration. The National Council of Women made this point in their evidence: 'the correct siting and relationship of each piece of apparatus is more important than expensive fittings in the elimination of unnecessary labour, fatigue and irritation born of frustration'.[2]

It was primarily for these reasons that the Dudley Committee deliberated at length on the suitable arrangement of kitchen, dining and living spaces. Pre-war practice, in local authority houses, was to have a small scullery with one or two living rooms. Meals would originally have been eaten in the living room. A coal range would also have been there, but when a gas or electric cooker was hired it would be placed in the scullery. Then the kitchen equipment would follow and people would take their mid-day meals there. It had been an intention of some designers to keep the scullery small in order to prevent meals from being eaten in it. This intention was presumably based on a desire to cultivate the appearance of a middle-class life style where the housewife would play the part of an unseen servant, so that meals would 'appear' in front of husband and children. The Dudley Committee regarded this attitude as wasteful, as the 1944 Housing Manual records: 'since the serving of meals in the most direct way possible from the cooking stove to the table greatly reduces the labour of housekeeping (CHAC 1944:17).

The Dudley Committee, in challenging pre-war social aspirations kept rigidly within the boundaries of the sexual division of labour. It was assumed that women's primary role was to service their husbands and children. For example, when presented with the following information about the timetable of a working-class household:

> The following time-table is not unusual in an average working household:
>
> | | |
> |---|---|
> | 7a.m. | Breakfast for Husband |
> | 8a.m. | Breakfast for Children |
> | 12.30p.m. | Lunch for Children |
> | 4.30 p.m. | Tea for Children |
> | 6.00 p.m. | Tea for Husband |
> | 7.00 p.m. | Supper for Children |
> | 9.00 p.m. | Supper for Husband |
>
> (CHAC 1944:13).

The Committee did not consider when the woman of the house might eat herself! Furthermore, no exploration was made of whether some more

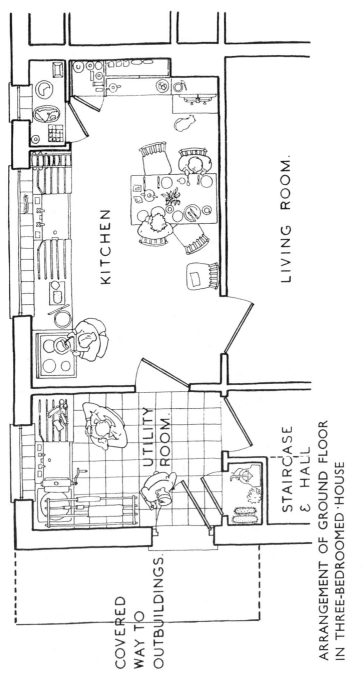

KITCHEN

LIVING ROOM.

UTILITY ROOM.

COVERED WAY TO OUTBUILDINGS.

STAIRCASE & HALL

ARRANGEMENT OF GROUND FLOOR
IN THREE-BEDROOMED·HOUSE

Figure 5.1 Alternative 1
Source: CHAC (1944)

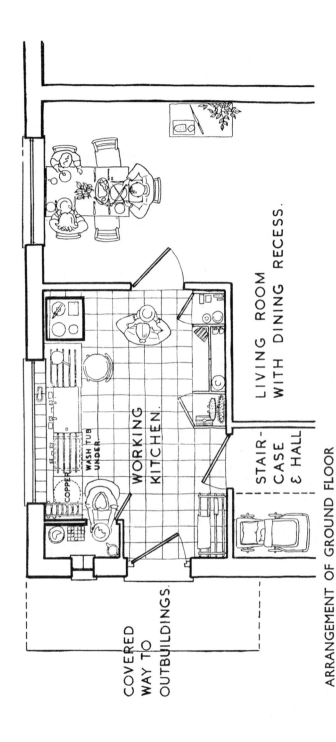

WORKING KITCHEN.

WASH TUB UNDER.

COPPER

STAIR-CASE & HALL

LIVING ROOM WITH DINING RECESS.

COVERED WAY TO OUTBUILDINGS.

ARRANGEMENT OF GROUND FLOOR
IN THREE-BEDROOMED HOUSE

Figure 5.2 Alternative 2
Source: CHAC (1944)

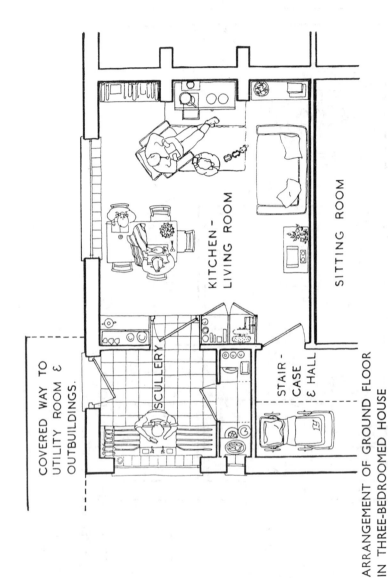

COVERED WAY TO
UTILITY ROOM &
OUTBUILDINGS.

SCULLERY

KITCHEN –
LIVING ROOM

SITTING ROOM

STAIR –
CASE
& HALL

ARRANGEMENT OF GROUND FLOOR
IN THREE-BEDROOMED HOUSE

Figure 5.3 Alternative 3
Source: CHAC (1944)

rational way of providing food might be considered, such as school meals or works canteens. Rather the Committee recommended three possible arrangements of scullery, kitchen, dining and living areas. These were presented in the Report of the Committee's findings (see Figs 5.1, 5.2 and 5.3).

The drawings of the three arrangements, which are reproduced from the original, illustrate prevailing assumptions about women's position. Alternative 2 (Fig. 5.2) provides the most extreme example with the husband carving and the wife serving the family from the working kitchen. In others, elder children appear to be helping with the washing up. Carving is however the most strenuous task the husband undertakes and in Alternative 3 (Fig. 5.3) he is portrayed by the fire with his feet up smoking his pipe.

Each example shows a tension between family togetherness and a desire to separate out certain domestic tasks. As a fundamental division each alternative has provision for washing clothes separated from eating. Alternative 2 (Fig. 5.2) is most like the pre-war semi-detached in plan: in this case, however, the kitchen is larger and has sufficient space for two people to work and for clothes to be dried. Alternative 3 (Fig. 5.3) also echoes the pre-war pattern of a by-law house with a living room with a coal range in it which, given the crockery and pots illustrated in the cupboards, appears to have been the sole means of cooking. This alternative appears to have been out-dated almost as it was published – as has been noted the sale of gas and electric cookers became widespread in houses with those services in the inter-war years.

Alternative 1 (Fig. 5.1) illustrates the forerunner of the modern dining kitchen. This 1944 model is, however, superior to subsequent arrangements made in the 1960s, '70s and '80s since it includes a utility room for washing and drying clothes and keeping brooms and other materials.

Thus the Committee answered the demands of women's organisations in its meticulous attention to the details of kitchen planning. It also included other demands from women's organisations in its Report and these were written into the Housing Manual. The demands were for piped hot water, more cupboard space, bathrooms and inside toilets, more shelving, better lighting and ventilation, power points, a higher standard of space heating, a larder, a sink, dry goods cupboards, kitchen storage, plate racks and drying racks.

However, it was in balancing priorities between space and equipment where the Committee took the lead from women's organisations. Both providing extra space and a higher standard of services and fittings in houses would add to the cost. The evidence from organisations such as the Women's Housing Advisory Council was that space should never be sacrificed to equipment. This point was emphasised to the Dudley Committee: '*Mean dimensions, once permitted, can never be remedied and are one of the basic factors in bad housing*' (emphasis in original).[3]

The Dudley Report recommended raising the size of the average three-bedroomed council house to 900 sq ft minimum. The Housing Manual was more cautious and recommended a lower standard of 800–900 sq ft. Through the influence of Aneurin Bevan, who was then Minister of Health, local authorities were persuaded to increase their space standards by 1,000 sq ft approximately by 1949. The 1949 Housing Manual supported this trend and proposed a standard of 900–950 sq ft. Before there was time to implement this fully the Conservatives won the 1951 election and Harold Macmillan became Minister of Housing. He recommended a lowering of space standards (Merrett 1979). In the period 1951–1957 space standards were sacrificed to a greater extent than fittings were. Although local authorities, such as the LCC, cut back on cupboard fittings, fires in bedrooms and other important but minor items, a reduction in floor areas and simplifying estate layouts were the main vehicles for a reduction in costs. Thus one of the most important elements in good housing design, in the view of the Women's Housing Advisory Council, was sacrificed.

The Dudley Committee was cautious in its embrace of technology. In North America central heating and fitted kitchens had been available for many years (Geidion 1948). Similarly, architects in Europe had been designing fitted kitchens and centrally heated houses. The Standing Joint Committee on Working Women's Associations, the Electrical and Gas Women's Associations, the RIBA and the Housing Centre all emphasised the need for a refrigerator. They argued that it was of equal importance to a cooker and necessary for reasons of hygiene, saving money through cutting down waste and saving labour in terms of reducing trips to the shops.[4]

In its Report the Dudley Committee had stated that it had considered the possibility of providing refrigerators and dishwashers. Whilst they hoped that their price would come down low enough to be within the reach of much of the population, they concluded that they did not find it practicable to provide them in municipal dwellings. Thus the Committee, whilst applauding modern technology in one part of the Report, drew back from a full acceptance of its possibilities. A similar timidity characterised the Committee's view towards central heating and argued that more research was necessary. Coal fires, with their attendant dirt and soot, were to be the order of the day. An opportunity to fully transform the home had been passed over.

The main reason for the Committee's caution was cost. Since they did not know what the arrangements for financing council housing were to be at the end of the war, they assumed that costs of construction would have to be paid for out of the rent. However, it is possible that had they recommended the provision of refrigerators, cookers and dishwashers, local authorities would have been able to order them in bulk, or manufacture them themselves and therefore reduce the unit cost. The same argument applies to central heating. By not recommending their provision, the Committee was perpetuating class differences in housing. Furthermore once the transition from coal range to cooker was complete, this meant that tenants then had to supply their own means of cooking when they took over a house.

The Dudley Committee, then, reduced class-status divisions in its recommendation of higher standards for council housing. Space standards for publicly owned and privately rented housing were beginning to merge. Figures for space standards of owner-occupied houses are sparse, but one study reveals that whereas in the 1920s 41% of houses bought by mortgage fell in the space range of most council houses (750–999 sq ft) by 1962 this figure had risen to 65%. Moreover, the number of houses built at the 'luxury' end of the range had fallen, by 1962 to an almost insignificant number. The difference in fixtures and fittings between local authority houses and cheap-to-medium priced private houses in the 1940s and the early part of the 1950s was minimal. Both lacked central heating but were supplied with back boilers, running hot water and electricity. By the latter end of the 1950s privately built houses might be offering more built-in appliances, such as refrigerators and clocks. However, the differences between public and private were small; these were amplified by builders offering a greater 'choice' to private buyers in terms of colour schemes, arrangements of rooms and minor changes in fixtures and fittings (Burnett 1980).

In this way the emergence of the 'housewife', a classless responsible citizen was aided. Whilst enormous class differences obviously still existed, the gap between the wives of manual workers and the wives of professional men narrowed. The former (once rehoused) no longer suffered a lack of basic utilities; the latter no longer had living-in servants. Differences of status had to be defined on new, more subtle lines, as glaring inequalities in the conditions of women's domestic work diminished.

It is worthy of mention, also, that in the brief period following the Second World War, space standards were higher than they had ever been and were ever to be again. Design standards had provided a battle ground for politicians in terms of subsidy. For example, Bevan, who was Minister of Health in the 1945–1951 Labour government, refused to allow Hugh Dalton, Chancellor of the Exchequer, to reduce space standards and therefore costs. Foot records how Bevan, with his Welsh mining background, was incensed by the cheeseparing attitudes of those from the upper classes who, as he said, would feel claustrophobic in a room 900 feet square, let alone a house of that size (Foot 1973).

Because so little is known about detailed design standards of owner-occupied houses, it is difficult to assess the impact of government space standards on the private sector. Certainly one study carried out in 1966 found that in a comparison of publicly and privately built houses, one quarter of the sample of eighty privately built dwellings fell below the then government standard (Burnett 1980). The assumption that privately owned middle-class housing provides a yardstick to which local authority housing has failed to aspire is erroneous. It is probably more likely that local authority housing provides some kind of standard for private housing – both in terms of trying to match it in levels of provision and in trying to distinguish itself from council housing in terms of style.[5] This question of style will be discussed again in Chapter seven.

Cleanliness and domestic appliances

In this discussion of housing standards it seems timely to mention some of the recent feminist debate on the connections between housework, gender and domestic appliances. In Chapter two (p.36) F.R.S. Yorke's belief that modern labour-saving devices would free women were referred to. This idea was commonplace until the 1970s, when Oakley (1976) and Vanek (1974) examined how much time women were actually spending in doing housework. They found that the introduction of household appliances in both Britain and the US had not made any difference to the hours which women spent doing housework. Vanek brought together a number of studies of housework hours in the United States from 1920 to 1960. She found that the number of hours which women who were full-time housewives, that is who were not employed outside the home, spent on housework barely changed from 1920 to 1960. In 1924 an average of 52 hours a week was spent on housework and in 1960, 55 hours per week. Women who were employed outside the home spent an average of 26 hours a week doing housework in the 1960s (Vanek 1974).

The reasons for this paradox were hypothesised by Vanek and supported by Davidoff (1976). She suggested that, whilst technical developments have meant that certain tasks such as food preparation may be done more quickly, changing circumstances have meant that other tasks such as household management have become more time consuming. Running a home involves more arrangement and co-ordination – of plumbers, officials from the statutory services and so on. Furthermore, changes in standards have also meant that more work is done to achieve a greater level of cleanliness. For example, time spent doing the laundry actually increased in her study period because people had more clothes and linen and washed them more often.

Vanek (1974) also suggested that non-employed women spent more time doing housework at the weekends than their employed counterparts. She suggested that this was because of the low-status of housework, such that non-employed women tended to do housework at weekends to demonstrate to their families that they were working during the week. This meant that their total housework hours became longer.

Bereano, Bose and Arnold examined the connection between housework hours and technology in greater detail. They found that the amount of time which wives in Britain spent on housework increased during the period 1900–1950, peaked in the 1950s and then suffered an absolute decline in the 1960s and 1970s. They broke down their statistics in terms of middle-class and working-class, part-time employed and non-employed women. Working-class women, part-time employed and non-employed women did more housework than their middle-class counterparts (Bereano, Bose & Arnold 1985).

The authors argue that the provision of (once) statutory services, that is, running water, gas, electricity and mains drainage affected housework time more than machinery or appliances used in housework. They cited a study made in 1929 which found that modern plumbing, electricity and equipment saved two hours a day in meal preparation and cleaning up and 0.7 hours for routine cleaning and the care of fires. They drew on Vanek's evidence and suggested that this saving in time was passed onto other domestic activities such as childcare, shopping and management. The suggestion that technology has 'liberated' women and freed them for waged work was refuted by the authors. They argued that the reduction in the size of families, which was particularly significant in Britain in the inter-war period was a more important factor.

This research emphasises the importance of raising housing standards to include the provision of easily maintained heating and running hot water. It is also interesting to note that it is the collectivised services, that is, gas, electricity and water, which have proved most labour saving and not individual appliances, despite the claims of manufacturers.

Forty suggests that the styling of domestic appliances and the claims made for them by manufacturers support an idea that housework is not real work. Firstly, appliances support the notion that housework is not work by taking over the work themselves. Secondly, by using references to full automation, a suggestion is made that housework takes no time. Finally, their appearance is designed to exclude any reference to paid work. Household machines do not look like machines used in a factory, or even an office. Indeed, the difference between, say, household domestic appliances and the type of machines which are used in commercial catering is quite striking, the one being elegant and crisp and the other an assemblage of moving parts. In this way divisions between work and home are emphasised (Forty 1977).

The raising of housework standards, which was made possible by higher standards of housing and smaller families, was supported by the domestic science movement. This movement, in Britain, appears to have been closely tied to the idea that women were subservient to men. Dyhouse describes how the Victorian middle classes wished to educate working-class girls to become 'good wives and little mothers' (Dyhouse 1981). There was an idea that poverty could be overcome by thrift and good household management. Dyhouse traces the history of the introduction of domestic science into the curriculum of elementary schools after the 1870 Education Act. There was an expansion of such education after the Boer War had revealed the extent of malnutrition amongst the working classes, for it was believed that such under-nourishment could be relieved by educational, rather than welfare measures.

Educational measures were adopted in the inter-war period to combat infant mortality. Ignorance rather than poverty was blamed for infant post-natal deaths caused by diarrhoea. Diarrhoea was attributed to dirty living conditions. Advice and education was given in schools for mothers and infant welfare centres, rather than more practical ameliorative measures such as free medical care or free meals. Medical officers of health and organisations concerned with infant welfare campaigned to have infant welfare and housecraft on the elementary school curriculum. Schools with a middle-class catchment disliked the term housecraft because it suggested domestic service. The subject was renamed with an increase in status to domestic science (Lewis 1980).

The domestic science movement pressed for a continual search for higher standards of housework. Although the term domestic science implies some line of logical thinking, of rationality, there was in fact a tension between the notion of rationality as applied to housework and ideas about hygiene and morality which were also associated with housework (Arnold & Burr 1985). In unravelling these I am not trying to suggest that housework is a futile activity, nor indeed that any feelings of pride in a clean house are evidence of false consciousness. There is a point to be made, however, between the need for cleanliness in terms of supporting life – for example, the demands of new-born babies and crawling toddlers are far in excess of adult requirements – and cultural norms and expectations.

In a previous section in this chapter commentators on slum dwellers' transition to new housing associated cleanliness with a rise through the class structure. Evidence of this attitude was found among housing managers in the 1960s (Tucker 1966). Furthermore, as just discussed the domestic science movement associated dirt with illness and, indeed, death. The belief that household dirt, as opposed to contaminated water supplies or food, or inadequate sewerage is connected to disease is difficult to sustain.

Ehrenreich and English undertook a survey of home economists to settle this point. They could find no connection between levels of cleaning and the incidence of disease (Ehrenreich & English 1979). In a similar vein Douglas could find no causal connection. The eradication of dirt, she argued, is a cultural activity, not a scientific, rational one. Household dirt is, she suggests, a misnomer. Many practices which are considered to be dirty, as, for example, putting shoes on tables or leaving underclothes in living rooms, putting plates on the floor are more to do with our notions of order than of disease-inducing lapses in hygiene. Cleaning and tidying are therefore ways of ordering the world (Douglas 1966). From a different perspective Forty argues that the styling of household appliances supports these notions of hygiene. He suggests that the persistence of the use of the colour white for their boxing is significant. Their design suggests cleanliness and that 'one might suppose that their function is to manufacture cleanliness' (Forty 1977:285).

Thus for a housewife to be dirty would suggest a challenging of class-status divisions, slovenliness, being 'bad' in an undefined sense and endangering her family's health. It is no wonder, then, that housework standards have been driven ever higher. The power of an ideology which identifies a 'good woman' as one who keeps a clean and tidy home is exemplified by Mackintosh's statement: 'The house is inseparable from the housewife. If it becomes dilapidated it becomes the wheel on which the housewife is broken' (Mackintosh 1952:110). As Comer pointed out twenty years later, in the 1970s, the housewife's self-respect hinged upon demonstrating to the outside world that housework is a moral activity and not a utilitarian one (Comer 1974). This imposes a heavy burden on women, for not only are tasks to be carried out, but they must be seen to have been carried out well. Such morality was not without its status connotation, as will be illustrated in Chapter seven.

So far only the level of fittings and design standards inside the home have been considered. In the next two sections attention will be turned to domestic tasks outside the home. In these instances I shall be using local evidence of examples from the London County Council (LCC) firstly, because no national evidence was readily available and secondly, because these examples have an immediacy which government reports lack – despite their occurrence some forty years ago.

Cleaning common areas

At a local level women could have power as councillors to raise housing standards, within the limits set by central government. One such woman was Evelyn Denington, who for many years was the Vice-Chair of the Housing Committee for the LCC. She made several interventions in order to raise standards for the benefit of women: her aim, as she stated in an interview in 1984, in terms of management as opposed to design, was that there should be sufficient council staff: 'Keeping the place nice, so even the Queen would not mind living there'.[6] Perhaps Denington's most interesting attempt to raise standards was her efforts to persuade the Council to take over the cleaning of common staircases in blocks of flats. This episode is significant in the perspective it throws on the value and costs of women's work and the type of work which 'housewives' were expected to do.

There had been complaints from tenants both during and after the Second World War about the requirement in their tenancy agreement to clean the common stairs. The normal practice was for the tenants on each landing to draw up a rota, with each (almost certainly) woman taking it in turns once a week to clean the landing and the flight of stairs to the next level. A petition presented to the Council in 1947 from tenants on the Millbank estate protested about the expense of the materials involved and the hardship endured by the many old people on the estate and on women 'who are out at work all day'.[7] The Valuer, who was the Council Officer who was then in charge of housing management, advised the Council not to change the cleaning arrangements on grounds of cost, arguing that difficulties occurred only in a small number of cases. The Council took his advice on that occasion.

In November 1952 the Council decided to hand over the cleaning of common staircases on all their estates to contractors and to charge three pence extra a week on the rent. This was done at Denington's insistence. As she said later: 'I wouldn't like to clean the stairs – it's disgusting. They are fouled by dogs and sometimes by children'.[8] Difficulties were soon experienced. The contractors did not do the job properly and tenants complained, 'not without justification in some areas' as the Valuer later reported. Accordingly, the Council advertised among its tenants for cleaners to come forward to do the work for the wage of 30 shillings a week. The response was poor, because: 'the work in winter is particularly hard as well as uncongenial when undertaken as a continuous task'.[9]

The Council decided to raise the specification and payment for the job. A sample survey was made of tenants on a number of estates to ask them whether they would prefer to pay 6 pence a week extra on the rent for the Council to clean the staircases. Of the 54,902 tenants asked, only 6,476 or 11% were in favour of this. As in only three small estates were there more than 50% of tenants in favour, the Director of Housing, who was now in charge of housing, advised that the experiment be dropped. His advice was taken.[10]

This episode raises some interesting points. Firstly, the resistance which both tenants and the Council put up to paying cash for cleaning illustrates the low esteem with which women's work was regarded. An inference to be drawn is that cleaning is regarded as a task which women would undertake as a matter of course and was therefore not 'work' and not worth paying for, whereas other routine maintenance tasks, such as replacing light bulbs was 'real' work and therefore worth employing somebody to do. Secondly, an attempt to limit the housewife's role to 'lighter' household work inside the home rather than 'heavy' and undoubtedly unpleasant tasks outside it had been thwarted. If Denington's plan had gone ahead, then the status of the work done inside the individual flat would have been raised as well as the calibre of the estates. This is because cleaning would have been demonstrably waged work outside the flats: therefore the inference that cleaning inside the flat also had a monetary value could not be ignored. That such a rise in status would have been justified is shown by the level of skill necessary to carry out the task effectively.

Communal facilities

Another way in which the burden of housework within the home could have been 'lightened' was by the provision of communal facilities or services. There were several different ways in which such services could have been organised which would have had different potentials.

Experimental groups of houses had been built at Letchworth and Welwyn Garden Cities and have already been described in Chapter two. Local authority-run restaurants during the war provided another example of a communal service (Roberts 1984). The emancipatory potential of communal services was three-fold. Firstly, they could relieve women of some household responsibilities. Secondly, they could provide a source of mutual sociability and support. Thirdly, they could challenge the privacy of the nuclear family, opening its most intimate transactions to outside inspection.

However, communal facilities could be provided for other purposes and be viewed by their users in a different light. Wilson suggests that the theme of home-making as a career and that of taking the drudgery out of housework were not mutually contradictory as might first appear (Wilson 1980). She argues that for homemaking to be made an attractive full-time career, some of the grind had to be taken out of housework, so that women could concentrate on the more stimulating and rewarding aspects of domestic life, such as childcare.

For example, the architect, Elizabeth Denby, in an article in *Picture Post* in 1941, proposed that after the war, laundry services should be extended. This was because she felt that washing laundry was too heavy a job for women to do as well as other household tasks and was better performed in factory conditions. Communal laundries were, she thought, inefficient, since they took women out of the home where they could have been engaged in other tasks. Women, she argued, were able to do several things at once: their time should not be wasted in a laundry for an hour or two every week, doing work that should be done by factory workers (Denby 1941).

What Denby called, the 'modern half-way compromise' of communal laundries had been provided by some municipalities before the Second World War. However, it seemed that use of their facilities carried some degree of stigma, as two surveys undertaken in 1939 and 1949 showed:

> Both the 1939 and 1949 surveys indicate that amongst 'the slightly better off housewives' (1949) a degree of social stigma was associated with using the public washhouse, particularly on Saturdays. According to one Bolton woman, Monday and Tuesday in the washhouse was quite popular but, 'only the bad type of woman washes later in the week. It shows she doesn't care. Especially Saturday is terrible.'
>
> (cited in Zmroczek 1984)

Thus washing, as a cultural activity, was not amenable to rationalisation.

The Dudley Report commented on the lack of communal facilities in flats. They were particularly enthusiastic about communal laundries, although some of the women's organisations had expressed reservations in their evidence. The Report stated that the Committee saw communal laundries as the best way forward for future provision, although they did make the proviso that more thought should be given to looking after small children whilst their mothers were using the laundry. A suggestion was made that laundries could be linked to creches or day nurseries where children could be supervised. The Committee did not state why they supported communal provision, but evidently thought that it would be beneficial as they suggested that tenants should be educated to use such facilities. The Ministry of Housing and Public Health was less enthusiastic and the Housing Manual 1944 merely noted that clothes-washing facilities should be provided as well as laundries.

The LCC had some commitment to the provision of socialised facilities. During the war the Council had run a communal meals service as part of the British Restaurants network and a small number of nursery schools. It continued to run some of the restaurants after the war and, after an initial setback, increased the number of nursery schools from five to twenty-five by 1963.

The Housing Committee made two attempts at running communal facilities for blocks of flats. The first was in childcare and the second was in the provision of nurseries. As will be seen, both schemes had been discontinued by the mid-1950s. The first had more emancipatory potential and was perhaps, for this reason, more tentative.

The failure of the first experiment in childcare was used to avoid further experimentation in this field. The original project was to provide playrooms at the Minerva Estate in Bethnal Green. The Save the Children Fund appointed a social worker, a Miss Davis to the estate. In April 1950 the Valuer reported:

> ...in some directions her work was very successful, but that she found great difficulty in obtaining the necessary assistance from mothers in the running of the playrooms – whilst they were content to accept the playroom facilities provided, they were not anxious to take their turn to supervise the rooms. The London Council of Social Services also report that there appeared to be a good deal of opposition from members of the estate Tenants Committee who, whilst their aims were the same as Miss Davis, appeared to have very different ideas as to how these should have been achieved.[11]

Miss Davis resigned soon afterwards. She was not replaced and eventually the playrooms were converted into flats.

The issue of supervised childcare was to arise again. An LCC conference on juvenile delinquency in September 1950 noted that a factor bearing on delinquency was, in their view, the number of mothers going out to work. A report was made in the proceedings of the Conference:

> We are far from willing to make a sweeping condemnation of all mothers, even of young children, who are employed.... We realise that employment is often an economic necessity for a hard pressed family; if this is so, we think that it calls for a review by the departments connected with social welfare.[12]

A suggestion was made that, among other things, supervised playrooms should be provided in blocks of flats, so that children whose parents had to be away, even for short periods, could be looked after.

The Valuer considered this request in 1951. He seems to have misunderstood its nature, for he suggested that the playrooms were needed for parents whose children were away for shopping, thereby trivialising the request. He went on to recall that the tenants in Minerva Street had been unwilling to organise a rota to supervise the playrooms themselves. His recommendation to the Housing Committee was that a scheme for providing playrooms with supervisors employed by the Council should not be proceeded with because it would be too expensive. As it was, he argued, the Council were already in deficit on its post-war schemes. The Housing Committee accepted his decision and the idea was dropped.

Unfortunately, the archives do not record the debates of LCC Committees, so that it is impossible to understand exactly why the idea was dropped. It would seem likely that the financial arguments were the most persuasive. The Council was already in debt over its post-war housing schemes, that is the building and land-acquisition costs were higher than the subsidies and expected rents; if the elected members of the Council had spent any more against the advice of their officers, they might have been liable to surcharge.

The Housing Committee was more concerned about the provision of laundries and the decision to discontinue them was not taken lightly. It would seem from an examination of the records that the laundries were provided, not so much with a view to liberating women from the confinement of domestic chores or to challenge gender stereotypes, but to lighten the burden of housework and to raise standards of home making. Thus the LCC's policy was more in accordance with dignifying housework as a job than with the establishment of a feminist prototype of an alternative society.

Immediately after the war the Valuer, who was also at that point Director of Housing, encouraged the building of laundries on estates. He thought that they:

> ...would not only remove the unsavoury features of the home
> laundry from the kitchens but would greatly lighten the burden of
> the housewives, and could be provided at a slight increase of cost
> which could reasonably met by an increase on the rent.[13]

Laundries were provided in inner-city sites. First, central laundries were built on estates then, when it was realised that these caused problems in terms of the distance over which it was necessary to walk to them, smaller laundries were constructed in different parts of each estate. The laundries were equipped with toddlers rooms which had glass screens so that the children could be supervised whilst the washing was being done. The layout of the laundry was also arranged so that each person washing could have some privacy. An extra charge was made on the rents for flats which had the use of laundries.[14]

Problems arose in 1950 because the Solicitor to the Council pointed out that, under the terms of the 1949 Housing Act, the Council could not charge tenants who did not use laundries for their use. However, an allowance for the fact that laundries were provided could be made in setting the rent.

Council officers were asked to report on the financial effects of providing laundries. A preliminary report was produced which suggested that the cost of building laundries on estates was two shillings and seven and a half pence per flat per week.[15]

The Finance Committee and the Controller of the Council, who had an overall view of the financial running of the Council, were dissatisfied with this state of affairs. They recommended that the laundries should be self-supporting and expressed concern that the Valuer had set the rents for some of the estates with communal laundries too low, only making an allowance of two shillings a week.[16]

The Director of Housing (there was, by now a separate post for Housing) revised his calculations and concluded that of the twenty-seven laundries which had been built, seventeen were in financial surplus and ten were in deficit. The average deficit overall was three-quarters of a penny per flat per week. This amount would have been reasonably substantial when multiplied by thousands of flats, per year. The Valuer also reported that he had allowed for longer operating hours 'to accommodate the needs of tenants whose wives were working'. This last phrase – the needs of tenants whose wives... – neatly encapsulates an assumption of male superiority and female invisibility which echoes through the decades since it was written! The figures for the use of each individual laundry ran from 24% to 100% of the tenants on each estate.[17] The Valuer thought that the reasons for non-use were that 'housewives' who were at work all day could not use the laundry and that in some cases the local bagwash – a service where clothes would be washed and roughly dried – was cheap. For example, a 12lb dry weight of washing would cost between two shillings and two shillings and sixpence a week at the bagwash.

The Finance Committee and the Controller kept up their pressure for a full report on the costs and use made of the Council's laundries. Finally, in 1952, a survey was made of 1300 tenants. Although the survey results do not state this explicitly, it seems that questions were asked not of the 'tenant', who was, as has been seen, regarded as the male breadwinner, but of wives of tenants. The evidence gained was ambiguous: 45% of tenants' wives with central laundries liked them and 58% of those with small laundries liked and used them. Where laundries were provided, approximately a quarter of women respondents did some washing in their flats and about a half used a commercial or bagwash service. A majority of women would have preferred to have had washing machines and driers in their own flats, provided by the Council and paid for on a meter; 44% on average would have preferred a wash-boiler and heated drying cabinet in their flat.[18] The Valuer did not make any recommendations, but on receipt of this evidence the Housing Committee decided not to build any more laundries on housing estates.[19] Those already built remained in use until they were all closed down in the 1970s.

The reasons why the programme for building laundries was dropped were not stated in the minutes. Presumably the Finance Committee put pressure on the Housing Committee to cut all possible costs from the housing budget. Arguments about equity between tenants who used and who did not use the laundries would probably also have been raised.

The LCC had not been committed to the more emancipatory aspects of providing communal facilities. Rather, emphasis had been placed on raising standards of housework and keeping heavy laundry out of the flats. For this reason no effort had been made to distinguish administratively between women who combined waged work and housework and full-time housewives. It could have been surmised that the former would have had different requirements from the latter in terms of time and money.

Rents and subsidies

Whilst a rise in housing standards might mean that working-class women could live and work in conditions that were also available to the wives of middle-class men, this rise was not achieved without a cost. Problems had already been experienced in the 1930s with families having such high housing costs that they were forced to cut back on food. Rowntree had found in a survey of York in 1935–1936 that 43% were in poverty. A summary of surveys carried out in the mid-1930s also found that approximately 10% of the country's population was undernourished: within this 10% were 20–25% of the nation's children (Marwick 1970).

Wages rose generally in real terms during the war (Addison 1977). In discussion in the army educational series about costs and standards there was a recognition that a rise in housing standards would mean a rise in costs and a rise in costs would have to be contained within the limits of working-class incomes. One bulletin put the choice starkly as a choice between higher rents or higher subsidies from central or local government (ABCA 1943a). Another added, poetically: 'It's no good planning a house on champagne lines for a mild-and-bitter income' (ABCA 1943b:4), thus neatly encapsulating class divisions.

Mass-Observation, in their survey for post-war hopes for housing, found a desire to see rents falling within a working-class budget (Mass-Observation 1943). The members of the Dudley Committee realised that their recommendations would mean higher costs. In the fluidity of war-time planning no real schemes for subsidies could be drawn up as there was uncertainty about post-war prices and future political directions in terms of state intervention.

In the detailed discussion of rent levels which follows I have drawn on central government and LCC policy. The experience of the LCC brought out not only the dilemmas about balancing rents, rates and subsidies, but also the assumptions on which rents were fixed. These assumptions were not spelled out in central government records of the period.

Experience from the First World War suggested that inflation in the price of building materials would cause prices to rise sharply after the Second World War. Bevan, Minister with responsibility for housing in

Atlee's Labour government after the Second World War, resolved to keep prices down by rejecting tenders that were too high.[20] He also managed to extract a greater proportion of Exchequer grant towards public housing.

Prior to the Second World War the ratio of Exchequer grant to rate fund contribution to the housing subsidy had been 2:1, that is for every pound that a local authority put towards the building of its council houses, central government would contribute two. Bevan managed to extract from the Treasury, against their advice, a ratio of 3:1. This ratio was kept intact by Macmillan, Minister of Housing in the subsequent Conservative government, again in spite of objections from the Treasury.[21]

Treasury objections were over rent levels. The Ministry of Health, that is, Bevan, wanted a subsidy which, on current estimates, would allow the rent on the average Council house in the provinces to be ten shillings a week. Rents for more expensive houses and for flats in London would be twelve shillings a week and for agricultural cottages seven shillings and six pence a week. This would have meant a rise of 40% on pre-war levels. Bevan thought that to charge any more than this would be 'asking for trouble'. Dalton, at the Treasury, argued that average rents should be higher, twelve shillings a week, on the basis that if a programme of building four million houses in ten years were to be achieved, it would not be possible to keep up such an increased level of central government subsidy. After some discussion Dalton gave way and rents were fixed at around the ten-shillings level.[22]

By 1948 local authorities were finding that they were having to charge more than the rent levels estimated in the Ministry's subsidy calculations to pay for increased loan charges and management and maintenance costs. Tenants associations, a trades council and local Labour parties wrote to the Ministry to complain.[23] Ministry officials reasoned that as the average gross rent of a new post-war council house was 14% of the average male income, there was no cause for alarm.

In all the discussions of rent levels, at central and local levels of government, the point for comparison was either the average male wage, or an average of male wages taken from some other source, such as waiting-list applicants. Occasionally the LCC considered household income, averaged out between household members. In these calculations, women's earnings were obscured. The setting of rent levels upheld the principle of the family wage.

The LCC experienced particular difficulties in relation to rents, prices and costs. Building costs were higher in London and land was more expensive. The 1946 Housing (Finance and Miscellaneous Provisions) Act kept the pre-war system whereby flats on expensive sites received extra subsidies. Even so, since many of the LCC's tenants earned less than the average national wage, and since the LCC was committed to providing a high standard of service, the tightrope between rents, subsidies and standards was often in danger of snapping. LCC flats were built to space standards below the minimum recommended by the Ministry. Although the Council had upgraded their pre-war designs in line with the Dudley Report, by 1949 they were having to cut standards on their cottage estates.[24]

By 1950 the whole subject came under question for the Council. The Brooke Report (Central Housing Advisory Committee 1952) had recommended that council houses should not be subsidised by more than the statutory contribution from the rates, so that the Council was unwilling to raise rates rather than rents. Eventually the rents were raised.[25] In 1953 a number of cuts in standards were considered and estimations of their costs made, but only a few were implemented.[26]

By 1955 the rents had to be raised again. The Valuer calculated that the average male wage for applicants on the waiting list was ten pounds, seventeen shillings and fivepence. The average gross weekly rent of all dwellings owned by the LCC was one pound, three shillings and fourpence. This meant that the ratio of average net rents to earnings was almost 11% – a comparatively low proportion.[27] For new post-war houses higher rents were charged. This was not because rents were fixed to the costs of construction for each block, but because the post-war blocks had better amenities, such as lifts, sun balconies and use of communal laundries.[28] In 1955 the rent for a three-bedroomed flat on a post-war estate was one pound, nine shillings and ninepence and one pound and four shillings for a two bedroomed flat. As the gross rent would have been half as much again the rent and rates would have been approximately two pounds, four shillings and eightpence and one pound and sixteen shillings – approximately 20% of an average income.

Whilst these rents might not have seemed excessive for those households dependent on an average income, where the male wage was below average, or where there were more than two children, or where there was a female head of household, these housing costs must have presented problems. A question was asked in full Council, but not given a reply because of lack of information, about how many offers of tenancies were refused because rents were too high.[29] Furthermore, it should be borne in mind that rents in the private sector, that is the greater proportion of rents, were lower. In the mid-1950s over 3 million dwellings were let at net weekly rents of ten shillings or less, and of these nearly one million were let at two shillings and sixpence or less (Cullingworth 1979).

For households coming into new local authority accommodation from the private rented sector this could have put pressure on their budget. The new flats, in some ways, possessed less fixtures and fittings than pre-war houses in that they were not provided with a range so that a cooker, and later a refrigerator had to be bought. As the floors were concrete, not wood, carpets were a necessity and not a luxury. A vacuum cleaner became more important. Furthermore 'moving up' into new housing raised expectations as to comfort, so that acquiring a three-piece suite, curtains and other furniture became more compelling.

These financial pressures contradicted the assumptions made in raising design standards. Whilst design standards were premised on the idea of women as 'respectable' housewives, who engaged in civic duties in the increased leisure time offered them by improved household equipment, the rise in prices and costs, plus the pressure to have an increased standard of living would have provided the impetus, together with the Government's employment policy, for women to join the labour force. These pressures would have been particularly acute in London.

Concluding comments

Housing conditions at the end of the Second World War were poor for a significant proportion of the population. The technology existed to produce domestic appliances on a large scale as in North America. Other methods of improving housing such as the provision of central heating had also been in use on the Continent and in North America for some years. Wartime had required the population to adapt to more communal forms of social organisation, both in civilian life and in the Forces. Therefore it seems, in hindsight, that the post-war period offered great opportunities to transform public housing.

Other influences, however, militated against dramatic change. Wartime communality had not led to a re-appraisal of woman's role of housewife and mother. Rather men's solidarity was called upon in opposition to femininity, calling upon men to forget their class differences and pull together to defend, among other things, their families. Such experiments in socialised domestic labour which had taken place during the war, such as the provision of restaurants and nurseries, were short-lived. Further experiments in socialised domestic work were hampered by an unwillingness to replace women's unpaid work with waged work and a lack of acknowledgement of the strain which women experienced in combining housework, waged employment and childcare.

A desire did exist, nevertheless, at central and local government levels, to lighten the burden of housework and to improve housing conditions. A government report was prepared which recommended an unprecedented increase in space standards: a recommendation which was wholeheartedly supported by women's organisations. Although significant improvements were recommended in fixtures and fittings, central government did not recommend the supply of central heating or modern domestic appliances such as cookers and washing machines.

In recommending and implementing increased standards for council housing, government was reducing class differences between working-class and middle-class households. In this period renting was the major tenure, so that class-status in housing terms was dependent upon type of housing, standard of provision and area. Given that housing improvements were predicated around women as housewives, women were the bearers of post-war egalitarianism.

Rising standards of housing meant also that standards of housework were raised. Domestic ideology was and is such that housework is not seen as a set of tasks, but as a moral undertaking. A woman's moral standing and status may be judged by the way in which she keeps her house and, by implication, cares for her family. In raising housing standards the material conditions of working-class women's domestic life were changed so that status divisions between working-class and middle-class households decreased. The absolute decline of domestic servants post-war reinforced this trend.

The Labour government took steps to ensure that subsidies for public housing were increased, so that rents could be paid for out of an average male working-class income. This meant that households who were headed by women would have found it difficult to gain access to council housing.

Despite extra subsidies, construction and land costs rose so that councils found it hard to keep rents to recommended levels. This, plus a rise in expectations of domestic comforts, which in turn had been stimulated by home improvements, provided an incentive for married women to take up waged employment. This was an acceleration of a pre-war trend. This was at odds with the ideal of the 'housewife' as a caring, responsible, house-proud woman with increased time for activities in the community.

A revolution in housing provision was not effected post-war. Pre-war trends were intensified in terms of raising standards of housework and in providing a stimulus for married women to engage in waged work. The structure of rents also promoted the formation of nuclear-family households headed by men. Significant changes were achieved in diminishing status divisions between working-class and middle-class homes. It was this change, plus the possibility of making significant architectural changes which fuelled architects' enthusiasm for public housing in the 1950s. It is this which I shall turn to next.

Chapter six

'We saw it as a dream'

The previous chapters have so far exposed the areas of continuity in housing provision and indeed in state policy. State housing was provided in ways which reinforced the values of the 'respectable' male working class: the superior position of the male breadwinner, the shaping of home as a retreat for men to return to and a rise in status achieved by improved standards of comfort and cleanliness. Set against this was a rise in housing costs and an increased acceptance, by planners at least, of the idea of married women returning to work.

Housing design had evolved slowly. The grim flats constructed by nineteenth-century housing trusts must have appeared an abrupt contrast with terraced housing: these had provided a model for local authority flats three decades later. The housing recommended by the Tudor Walters Report had been a decisive break with by-law housing in terms of its style, form and content: however it was still two-storey, low-density housing. In the inter-war period council housing was mainly composed of two types – walk-up blocks in the inner city and two-storey cottage estates in the suburbs. Occasionally a council might do an experimental scheme, of which Quarry Hill flats in Leeds provides a prime example (Ravetz 1974).

In Europe new styles of architecture had been developing from the beginning of the twentieth century. Although their proponents claimed that theirs was not a style but a rigorous application of technology to a building programme, the evidence of subsequent historians has explained clearly that modernism was a style, as well as a philosophy (Banham 1960; Jenks 1973). In modernism's international phase, which it entered in the 1930s, the style could be recognised superficially by its adherence to the use of steel-and-concrete framed buildings, lack of ornament or decoration, flat roofs, 'pilotis' and horizontal bands of windows.

Whilst the aesthetics of the style were somewhat uniform, its best practitioners escaped monotony by an imaginative use of sculptural space and an intensive examination of the building's purpose.[1] In each project the use which the building would be put to was analysed and each function assigned a separate part of the building or form. These parts would then be assembled to form a coherent whole which might not, in its shape, relate to any previous morphology which had been previously used for that building type. For this reason modernist architects were interested in a building's use and its part in a higher ideal of the organisation of society.

Modernism, then, offered up the possibility of a completely new approach to design. This chapter will examine the approach of a modernist group of architects to council housing in the post-war period. As ever, tensions of class and gender form a context to their deliberations.

The LCC Architects Department

In 1956 the Architects Department of the London County Council was described as being 'the most successful architectural office in the world' by Robert Furneaux Jordan, himself a distinguished architectural journalist. He thought that the virtues of the department lay not only in its aesthetic and technological achievements, but also in its social intentions. He thought that the architecture produced by the LCC related carefully not only to the Greater London Plan but to the 'social objectives of enlightened and humane government' (Furneaux Jordan 1956:304). This praise of the Department at this time, although extreme, was not unusual.

The housing division of the Architects Department in the LCC had been in existence since 1893. Prior to the First World War, it produced some flats and cottage estates which won acclaim and recognition from within the profession (Beattie 1980). In the inter-war period the Department's quality of work declined in freshness and vitality: at the outbreak of war their output was confined to neo-Georgian flatted estates in the inner areas and cottage estates in the suburbs. The Chief Architect, Topham Forrest had insisted on a neo-Georgian style in order to ensure a domestic appearance to the flats. His attempts to experiment with new housing forms, such as nine-storey flats, led to criticisms from proponents of garden cities within the architectural profession and he was generally constrained by the structure of subsidies and land prices (Pepper 1981).

After the Second World War the LCC faced a severe housing shortage. The officers estimated that dwellings for 400,000 people had to be built in the first four years following the German surrender.[2] The Valuers Department, which was staffed by surveyors, was given the responsibility for all the new housing work. Houses and flats were designed and built on pre-war lines with improved fixtures and fittings as outlined in the *Housing Manual* (MoH 1944). Estates of two-storey houses were built out-county and four-storey blocks of flats in-county.

Although the Valuers Department achieved a high level of output, the quality of their work was criticised. In the early part of 1949 these criticisms were made public, in a radio broadcast and in the architectural press.[3] The attacks were based on aesthetic grounds, that the quality of housing was dull and uniform and that the natural features of the sites, in particular the trees, were being cleared.

Evelyn Denington, who was Vice-Chair of the Housing Committee and Chair of its two sub-committees, supported the criticisms, and the view of the architects, although she could appreciate the Valuer's attempts to re-house the maximum number of people possible.[4] At the end of 1949, control of new housing design was transferred to the new Architects Department. Their staff was expanded and the Council asked them to produce schemes of 'the highest possible quality'.[5] A new housing section was formed in 1950. Whilst the division was subject to the overall control of the Council's architect, Robert Matthew, the Deputy Architect, J. Leslie Martin had responsibility for the general policy of architectural development within the Division. Martin was already recognised as an impressive designer through some of his early designs for the Royal Festival Hall. Although having firm ideas on design himself, he allowed different schools of thought to develop among the young architects in his Housing Division.

There is evidence that a significant proportion of the architects within the Department were either members of or close to the Communist Party. Banham (1966) suggests that this led to strong debates within the division about style, with a form of socialist realism being proposed by the Communists and sympathisers. Whatever the truth of the matter in terms of the political differences within the Department, there is much evidence to suggest that the standard of recruits to the Department was high and that their level of enthusiasm was above average. As the following statements from two of the new entrants show, their interest was not only in architectural appearance but in the purpose of the buildings themselves and the contribution to society which they could make:

> ...everybody came together ... to the local authority offices, which were the new sort of vision, it was the Welfare State, it was the thing that everybody was conscious of, and I think very firmly *believed* in, there were very few dissidents.... Everybody had a gilded vision of what life would be like for everybody living in this new society. Everybody was conscious that it should be a sort of beautiful society.[6]

> There was a euphoria about in the Housing Division and there was an enthusiasm. Most architects offices are very enthusiastic about their work ... but I have never met quite the level of enthusiasm that there was in the offices of the LCC. There were very vital discussions and electrifying debates and an intensity of effort that

was concerned very much with the quality of architecture and the quality of service to the community that architecture could give.

(Partridge 1980:170)

Mixed development

The major architectural innovation which the architects introduced was the concept of mixed development. Bevan's idea of mixed development has already been discussed in Chapter three, that is the notion of mixed development as a mixture of social classes. Mixed development could also be defined in an economic sense by having an estate with differential rents for different types of dwelling. Mixed development in an architectural sense had a different meaning. The LCC Research and Development Officer, Cleeve Barr, defined it as a variety of types of dwelling inhabited, as far as was possible by a variety of types of family which would give a lively and differentiated appearance to a housing scheme (Cleeve Barr 1958).

Leslie Martin favoured mixed development as he saw it as a way of breaking out of the straitjacket of designing high-density flats in the inner city and low-density housing in the suburbs. By changing to mixed development a new series of architectural questions could be posed: how were the dwellings to be arranged, in what groupings? How much privacy should each have? What should be the balance of private space and public open space? What kind of environment should be created? How urban should it be? (Martin 1983). These questions and more became matters of debate within the architectural profession, debates which were to continue into the 1970s.

The LCC did pursue the idea of mixed development in a social sense, by building some houses on particular estates to a higher standard and charging higher rents for them. In this way the Council hoped to attract professional middle-class people on to their estates. These experiments were limited in scope: the architects' interest in class revolving more around aesthetic considerations, as will be seen.

Mixed development and community

In a joint report to a government sub-committee the Architect to the Council and the Director of Housing both stated that they wished to create the appearance of a community:

In connection with a scheme of development of high flats, it must be borne in mind that the aim is to create homes and that, generally speaking, the old idea of the village or communal settlement is kept well to the forefront.[7]

That the image of a village community should be invoked in connection with a new housing development is not surprising. As early as 1901, the architect Raymond Unwin wrote of the need to impart a sense of community to council estates. He proposed to do this by calling on the imagery of a rural village (Unwin & Parker 1901). The ideal of communal settlement was pursued in the design of some inter-war council estates, for example, Bentley has argued that the key difference between publicly provided and speculatively built housing estates in the inter-war period was the notion of collectivity. Local authority estates were designed to evoke an effect of 'community'. This would be achieved in formal aesthetic terms by arranging the houses around a communal green space so that visually the collection of individual houses had some unity. Speculatively built estates, by contrast emphasised the individuality and privacy of each dwelling (Oliver, Davis & Bentley 1981).

The word community or communal has a rich complexity of meaning. Newby (1980) has identified three different meanings which are associated with the rhetoric of village community. The first is simply a geographical expression using the word to refer to a settlement of some kind. The second is of a pattern of social relationships, with the idea that a community can be said to exist when everybody knew everybody else. The third meaning is that of the 'spirit' of community, a sense of belonging, a shared social identity. This last nebulous sense refers to a subjective experience which is more commonly believed to occur in villages. Architects, by building in a vernacular style in groups of cottages which evoke a village visually seem to have desired the third effect or 'spirit' to arise as a corollary.

Davidoff, L'Esperance and Newby (1976) have commented on the persistence culturally of the idea that a rural village provides the ideal setting for a home. They have suggested, through an examination of literature and historical precedent, that although a romantic idealism evokes village life as full of contentment, harmony and peace, conflict and oppression were more likely to have been the reality of social relations in rural life. Women were subordinated to men, masters to servants and labourers to squires. Patronage rather than democracy ensured quietude. Patronage within the context of the village extended from the squire or local land owners to the villagers or landless labourers and within the squire's house extended from the master of the house to the female relatives and to the male and female servants.

Newby has also argued that the village in the form in which it is commonly understood as a nucleated settlement only dates in lowland England from the beginning of the industrial revolution. The village is neither as unchanging nor as timeless as is commonly supposed (Newby 1980).

Williams (1975) has suggested that the type of community which village life promotes may not be as tight knit as is commonly depicted. He questions whether a sense of community existed in rural settlements before the enclosures of the seventeenth century. Communal feelings, Williams argued, tended to be based on shared feelings of oppression, against an employer or a landlord or through an external threat and generally were recorded much later in time. Newby similarly argues that in the post-war twentieth-century village the existing community of farm workers was united through a common experience of employment and shared sense of relative deprivation. Newcomers to the village could not partake of the shared set of values which the farm workers held and disrupted that communality.

Thus, the communal life that was enjoyed in villages seems to have been based more on the social circumstances of their inhabitants rather than being derived from the picturesque formal qualities of the buildings. Yet this is not to say that there is no relation between the shape and arrangement of buildings and the social life that goes on within them. Rather, that the social relations are not magically imparted from groups of buildings just because architects wish it. Furthermore, certain aspects of behaviour may change, or be made more difficult as the next section will demonstrate.

Landscapes and domestic work

The origins of the rural village as a picturesque nucleated settlement may be traced back to the enclosures of the sixteenth to eighteenth centuries. In what has been termed the agrarian revolution, prosperous land owners enclosed common lands and amalgamated their estates to form large agricultural holdings. With the surplus generated by these proceedings the land owners were able to build themselves prestigious country houses surrounded by landscaped parks. Williams (1975) comments that in the creation of this new form of nature, the landscape was manipulated so that any elements of production, such as farm buildings, animals or farm labourers were removed. There were instances where entire villages were erased from parks to be rebuilt outside the park gates. In improving the landscape some land owners built model villages for their workers, outside the gates and railing which protected their country estates (Darley 1975).

From Elizabethan times there were two sorts of villages, the 'open' and the 'closed'. The closed village was owned by one land owner who could regulate who could live there and under what conditions. The 'open' villages were unregulated and anyone could live there. It was the 'closed' villages which tended to be most visually appealing, since the land owner had aesthetic control. These villages suffered most from the paternalistic control of the land owner since his displeasure could deprive farm labourer of both home and job. Yet these villages were also exclusive by nature and tended to have better housing. It is not surprising that the first suburban council houses, built for the more well-paid sections of the working class and managed paternalistically, should have been modelled on the pattern of the 'closed' village.[8]

In forging a new imagery for their post-war in-County estates, the LCC architects jumped over the park gates as it were. Estates were landscaped in the manner of aristocratic parks. Perhaps the prime example of this is the LCC's most famous estate architecturally, Roehampton. In both Alton East and Alton West (both halves of the same scheme) tall blocks of flats were set in rolling green countryside. The existing trees were retained and smooth expanses of clipped lawns were provided.

Alton West was particularly striking (see Figure 6.1). Its main feature was five eleven-storey slab blocks. These had been influenced by Le Corbusier's designs, in their position as free-standing objects on a smooth green lawn supported on stilts or 'pilotis' and with their crisp geometry of balconies and pre-cast concrete panels. The blocks have a lyrical quality, evocative of Le Corbusier's villa at Poissy which seems to float above the surrounding countryside.

Figure 6.1   Alton West

Figure 6.2   An ordered country house

The inspiration for this type of informal yet mannered landscape was provided in the pages of the journal *Architectural Review*:[9] help also came from the sites themselves. The Roehampton estates were set in some gardens which had been originally laid out by Paxton himself and on the Crescent Estate, which will be examined in the next chapter, the flats had been built in the grounds of some large Victorian houses. Another source for Roehampton could also be traced back to Le Corbusier's design for a city of three million inhabitants, drawn up in 1925 (Le Corbusier 1929).

Just as in the park of the country house, any suggestion of agricultural production was suppressed (see Figure 6.2), in these new LCC estates visual signs of domestic production were also organised or eliminated. The architects saw these signs as elements of disorder, which distracted from the overall harmony of the scheme. Great care was taken that disorderly features such as washing, should be hidden from general view. For example, in a committee report which discussed whether to provide the relatively expensive item of solid walling to some gardens, the officers commented:

> ...any untidiness would be made less apparent by some form of solid screen: in addition drying of washing will certainly be done in the

gardens, though this may occur only once a week, and unless it is screened it will be obvious from other parts of the estate.[10]

Small gardens and washing lines were provided to maisonettes, but these were situated behind the blocks. Stores were provided for tenants' prams and other household items. There were no private gardens around the flat blocks themselves. These were felt to conflict with the overriding aim of opening up views of the landscape. Private 'backyard' gardens, with their clothes posts and rabbit hutches, were thought to be 'untidy by their very nature' (Cleeve Barr 1958:37). Children's play was confined to organised play areas, fenced and walled-in. As will be seen in the next chapter there were rules governing the hanging out of washing in the flat blocks. On the Crescent Estate, the flat blocks were approached by winding paths, just as in the aristocrat's landscaped park, and roads were hidden by trees.

The visual metaphor of the landscaped park draws on the vocabulary of a class superior to the rural peasantry housed in the village. This was an effect not unwelcomed by the architects as one explained, with reference to Alton West:

> ...some people saw the vision of everybody living in a parkland setting, an Elysian parkland setting, the white modern buildings that had no reference to anything that had happened before, set in this idyllic field, with beautiful cut lawns ... there was an idea, and it was a middle-class idea curiously enough, do you see, everybody was trying to make everybody live like they thought they would like to live.
> [Could you describe the nature of the middle-classness of it?]
> I think the nature of the middle classes is a sort of nouveau situation, which is often a pleasant one ... if you're brought up with a garden with flowers in it, your parents liked those flowers and that garden, and there was sort of, you know, the click of the cricket bat and the cooing of the wood pigeons and things like that. There was that sort of idea which everybody thought so nice ... particularly, it always related to ... Oxbridge because quite a lot of those students had been to Oxbridge ... and there were marvellous buildings set in lawns and gardens.[11]

In designing these new post-war estates the architects were able to skilfully combine new forms of dwelling such as point blocks and T-shaped blocks of flats and maisonettes, in schemes which drew on older, more traditional forms of landscape design. This re-ordering of old and new elements rather than transcending class was evocative of a middle-class way of life in its visual order and harmony.

Concluding comments

Thus in their commitment to building for a new society, the architects drew on their own dreams and aspirations. The architect for the Crescent Estate responded to a question about the kind of people whom he imagined moving onto the LCC's new estates as follows:

> ...it would not be housing for the poor and it would not be housing for the underprivileged. It would simply be housing on a rental basis. And we saw it as anybody moving into these houses. And if I say that I suppose one had a middle-class idea of this it wasn't designed for the middle classes, we thought it would be nice to have privileges which maybe we had enjoyed and others hadn't which would be universally available to everyone. We saw it as a dream.[12]

In their implementation of their dream the architects of the Housing Division drew on rural imagery, on what Davidoff, L'Esperance and Newby call the 'country of the mind'. Unlike Unwin, and the architects who followed him in the inter-war period, they looked not to the imagery of the 'closed' rural village for their model but to its aristocratic counterpart, the country park. In this preoccupation with class and representations of class, issues of the actuality of working women's lives were avoided.

Just as the eighteenth-century land owner would look out at the view from the loggia of his country seat, so the tenants on the LCC's new council housing estates could gaze at the 'pleasing prospects' of their estate from the windows of their flats. Just as intrusive elements, such as farm labourers, ploughs and barns were cleared from the park, so were the clutter of domesticity, rabbit hutches, dustbins and working life banished from council housing estates.

Whereas an eighteenth-century land owner could rely on hidden hands, an army of servants, to perform unseen his domestic chores, council tenants had only themselves to rely on, and the services of the Council itself. It was the tenants, and wives in particular, who had to look after their children, do their own washing, cleaning and tidying, as well as earning a living. How some of these tenants fared, how they perceived themselves and how they saw the 'dream' will be explored in the next chapter.

Chapter seven

A respectable life

The past chapters have shown how housing and planning policies were shaped towards maintaining an ideal of a family unit in which women were the primary home makers and child rearers. At the same time these policies were framed in the context of a capitalist economic system where state intervention could only mediate the effect of market forces rather than change them fundamentally. This meant that for the majority of people, housing was provided according to ability to pay rather than by need. Council housing was allocated according to need, but only for a limited number of households and only for those households which accorded to a norm set by policy makers. Because council housing was produced by the state competing with private capital for land, council houses were produced, in inner urban areas, mainly in a form which was not conducive to the bringing up of children.

In this period following the Second World War, the majority of households rented their accommodation from a landlord. There was no stigma attached to renting. However, the point at issue was the type of accommodation and the standard to which it was kept. Differences in class-status revolved around these refinements. In the service flats of Mayfair and St Johns Wood in the inter-war period servants kept entrances and halls clean and fragrant and property companies ensured that the property was well maintained. This was the standard which households could aspire to. Furthermore, at the top of the tree so to speak, were the great houses of the aristocracy with their army of servants, perfect interiors and exquisite landscapes.

House and home as physical and social structures provide an exact interface of class-status and gender systems. A household's wealth is expressed in its home. At the same time a woman's worth as a good wife, a good mother is demonstrated in the keeping of house and home. Previous feminist studies have tended to emphasise the common experience of women, as women. It is the argument of this chapter that the divisions between women, which are expressed through home making, are important in terms of an external class-status system. These divisions were brought to the fore in my thesis study of an estate built during the early 1950s by the LCC (Roberts 1986). Eighteen people were interviewed, fifteen men and three women: all had moved onto the estate when it first opened and had lived there ever since. The interviews were semi-structured and lasted on average for one hour and a half. Respondents were asked to talk about the estate when it first opened, rather than how it is now. The interviewees were chosen by a variety of methods, so as to ensure a cross-section of opinion. Although there were differences between tenants, there were also similarities and it is these which will be examined first.

'The best estate in London'

The Crescent Estate was chosen for its relative ordinariness, as the subject for a case study. The post-war LCC estates which have received most publicity and the greatest amount of attention from social scientists, are those built in Roehampton, that is Alton East, Alton West and the Ackroyden. The Crescent Estate had not received any prizes: it is mentioned in an official history of LCC architecture as being 'traditional' in its design and therefore, by implication, uninteresting. The estate is of mixed development: there are two-storey houses, four-storey terraces of maisonettes, three-storey blocks of flats and five-storey blocks of flats. All are constructed with brick cladding and the houses have pitched roofs. As such, the construction is within the mainstream of British public housing. The estate is situated in North London, in an area which, although it is fashionable now, was working class when the estate was built. The architect who was the 'job' architect for the Crescent referred to it as 'this rather ordinary housing estate' in an interview.[1]

Yet the majority of the tenants interviewed felt the estate to be superior. In a few instances this sense of superiority rose to the level of hyperbole, as is illustrated by the following comments:

People were jealous of us when we first moved here.

(Miss Young)

Soon after this estate was finished we had coach loads of students ... Russians ... people to come and inspect the estate because it was one of the GLC's show estates.

(Mrs Sugden)

Well, to me there's no estate that compares with this estate.

(Mrs Goody)

Figure 7.1 Crescent Estate: site plan

But when they were first up, you know, they used to bring coach
loads of Russians round to look, this was the prize estate of the
GLC.... You could travel the whole of London and you wouldn't
find a flat like this.

(Miss Young)[2]

It seems quite likely that the estate was visited both by students and by
Russian visitors. The minutes of the LCC Housing Committee record
many visits by foreign delegations but no details were kept of which estates they visited. That the estate was a showpiece is open to disbelief.
The 'prize' estates in this period were Alton East and Ackroyden. Indeed,
the Ackroyden won a medal from the Royal Institute of British Architects
for good design. Claims that the Crescent Estate was the best estate in
London have no foundation in the pecking order as recorded by architectural journals nor in the minutes of the LCC Housing Committee. Yet this
representation of the estate was voiced by a number of interviewees.

One factor in the estate's assumed superiority was its location. The
Crescent Estate is situated at the far end of a private estate which was built
in the nineteenth century. Two of the roads onto it, Lowfield Terrace and
Lowfield Vale were laid out in the nineteenth century. Lowfield Vale is
a long road, at least a mile in length, with the Crescent Estate at one end
and the original Victorian houses along its length. Lowfield Vale and its
associated streets were laid out in the early years of the nineteenth century. In her history of the area Hinchcliffe (1981) describes how
prosperous merchants moved out from the City to the new suburb, using
a recently built railway station to travel to work. They lived in large detached mansions in a parkland setting. Each house had servants, and some
had coach houses.

From the end of the nineteenth century the area had been in decline.
Lowfield Vale became surrounded by new development as London
spread. The houses were let into multiple occupation and two council
estates were built nearby.

Figure 7.1 shows a site plan of the estate. The estate is bounded by two
major roads. One leads to the tube station some half a mile away. This
area experienced riots in 1981 and has an unpleasant reputation for viol-
ent crime. As an area it seems to have been regarded as 'rough' working
class through its history (White 1986).

The area surrounding the estate was capable, in the 1950s, of being
classified in either of two ways. It could either be considered as part of
the area around the tube station, and therefore undesirable, or it could be
regarded as part of Lowfield Vale, and hence desirable, particularly when
the road's past glories were remembered.

One man associated the estate with the area around the tube station,
which he disliked.

Actually I'm not all that keen on the location. That area, as soon as
you move out of the estate ... I remember it going back as a lad, it
was always an unpleasant grotty sort of area. And I was conscious
of that.

(Mr Nash)

Six others were positive in their evaluation of the area. One woman said
simply that it was a better area than the estate she used to live in, which
was Woodberry Down in Hackney. The others identified strongly with
the past history of Lowfield Vale.

It was a first-class district then.... Mother can remember this area
when there were carriages driving down Lowfield Vale with
dalmatians trotting behind them.

(Miss Young)

These sentiments were echoed in a conversation recorded between two
interviewees.

Mrs Sugden: And the area, as far as it goes, it's still pretty good.
Mr Sugden: And it still is, oddly enough. And in fact Lowfield Vale
many years ago was a private road.... Yes, there were great big
houses along there, as I say it still is a nice area. Fifty years ago
there were three millionaires that lived in Lowfield Vale. They've
still got the gates, the gates at the bottom here.

One woman even found the area more interesting to talk about than the
estate.

The thing that was interesting here were the lovely old houses they
pulled down to build those wretched flats. That's more interesting....
It was beautiful, I mean I can remember coming up here as a kid,
lovely old houses. You see, by all accounts, you take this Lowfield
Vale, they were full of these beautiful old houses, big houses,
seventeen-, eighteen-room houses, they all had their servants and
they all had their horse and carriage.

(Miss Read)

To conceptualise the area in which one lives as being of high status, by
virtue of its proximity to an area in which upper-class people live (or once
lived) is unexceptionable in middle-class behaviour. It is an attitude
which estate agents exploit. Although this kind of snobbishness is commonplace amongst owner occupiers, previous studies of Council tenants
have not suggested that they shared in this particular aspect of status.

Figure 7.2   Interior: Crescent Estate

Next to the estate, on one side of Lowfield Vale, is another small group of Council flats. These were constructed to the LCC's previous design, with small access galleries. The colour of the brickwork is similar to the five-storey blocks on the Crescent, but otherwise the designs are quite different. However, they were both managed by the same landlord. It is significant that none of the interviewees mentioned this estate directly. Either vague illusions were made to 'other estates near here' or more precise references to some tower blocks a quarter of a mile away. It was as if this other small, neighbouring Council estate did not exist: it had not acquired the status of being 'one of us'.

Twelve of the respondents expressed enthusiasm for the Crescent Estate itself. Different facets of the estate were emphasised: but each had a common theme of the superiority of the estate. The design and layout provided the basis for much of the praise. One woman drew an analogy between the estate and a private estate:

I've always had thought that this estate in its layout equates very well with a private estate. The layout of this I cannot fault.
[How do you see it being like a private estate?]
In the way we're not all concrete are we? I mean whichever way I look out of my flat, that way I look on a square, with green and

trees, that way I look on a park, that way I look on another square as you can see. Now in not many council estates can you do that (see Figure 7.2).

(Mrs Ivel)

This tenant's mother, who lived in the same flat, had suggested to me that the estate had been built for private tenants first, 'then the Council took it over'. This view was erroneous but shows how powerful her feeling of superiority was.

The idea that greenery and council estates were an exceptional combination was echoed by three tenants. One man thought that the lack of trees had previously been the hall-mark of local authority inner-city housing:

One big thing that you notice when we first come here was the way that they preserved all the trees; it seemed that they went out of the way to keep the trees going, whereas years ago, before the war, council blocks were like barracks. Asphalt, and you knew you were living in council flats.

(Mr Nash)

The grass and the trees were valued highly by eight of the tenants. Comparisons were made between the appearance of the estate and the 'feel'

Figure 7.3   Houses and low-rise flats: Crescent Estate

of the countryside. Three tenants compared the country atmosphere to that archetypal image, the rural village.

> ...when my mother was alive, she wasn't very well, I would have her to stay and she used to say to me, 'it's like being in the country, you don't see anybody'. The people go off to work, children go to school and in the summer months it's lovely here. It's like a little village.
>
> (Mrs Goody)

In the case of the Crescent Estate the image of a rural village acted at the level of aesthetic sensibility only (see Figure 7.3). The Crescent had a relatively high density of 100 people per acre, unlike the pre-war suburban estates which had densities of approximately 35 people to the acre. The estate was near to shops and centres of employment and women were not restricted by lack of access to transport. This is illustrated by the fact that of the twelve women interviewed, six had outside employment when they moved onto the estate.

Of the tenants who valued greenery, four suggested that the landscaping itself contributed to the unique character of the estate, and hence to its pre-eminence, not only over other LCC estates, but in the whole of London.

> You couldn't ask for a better layout, with windows all around. Wherever you are you can see greenery.
>
> (Miss Young)

> There are very few estates, are there, where you could go round and see rose trees, can you, and still blooming after twenty-eight years?
>
> (Mrs Taylor)

> Yes, it was, it was one of the nicest estates I've ever seen in London. It is and was and whoever planned it, they left all the lovely trees and scenery they could, they left it there.
>
> (Mrs Leary)

> How it's laid out is really, I think lovely. I don't really, truthfully, don't think there's another estate (to compare to this) ... I mean they might be better inside, they're more modern and got everything, you know?
>
> (Mrs Goody)

Another feature which may have raised its status from the tenants' point of view was that the estate was built opposite a park (see Figure 7.1).

Figure 7.4 Greenery: Crescent Estate

Two tenants referred specifically to it as an asset and described how much they enjoyed sitting and walking in it. Undoubtedly the view onto the park, which has some fine, mature trees, contributed to the general impression of greenery to which the tenants ascribed a high status. Young and Wilmott (1973) have commented on the way that overlooking a green space is a privilege that the rich have procured in London – Richmond Park, Holland Park and Dulwich Common providing key examples. Even in poorer areas, house values and rents rise near the park, as with Victoria Park in Hackney. The siting of the Crescent Estate must have contributed to perceptions of the estate's superiority.

As well as the greenery, the layout of the estate was appreciated by three of the tenants. The bulk of the accommodation was in five-storey T-shaped flat blocks. The way in which they were arranged, so that they all skewed away from each other and presented a varied appearance, was admired. Again comparisons were made with the 'barrack'-like appearance of pre-war council estates.

> So really, we always say we're just proud of it around here ... and I think it's the way the estate is laid out that gives people a pride in it. It must be something to do with how the place is being built – you

get rows and rows of barrack-like blocks, I don't think people take
so much pride in it.

(Mrs Goody)

Management

The LCC's management of the estate when it first opened supported notions of the superiority of the estate. Because council housing is a scarce
and desirable resource, a selection of tenants has to be made. The LCC's
point system has been explained in a previous chapter: the points system
was established to give priority to certain categories of need. Over and
above this more refined methods of selection were used to allocate particular flats on particular estates to certain types of tenant.

Although this further selection was not recorded formally in any
minutes, reports of the allocation procedures of other authorities in the
1950s, 60s and 70s have shown how selection was based on housekeeping
standards and other criteria which included the personal prejudices of the
housing visitor (Ginsburg 1979; Power 1987). Certainly the rent book and
records of a model LCC tenant kept in the archives include comments by
a housing officer on the tidiness of the tenant's house.[3] New-built estates
are sought after in any authority and it is in allocation to these that housing
officers can exert the most discretion.

That this choice operated in such a way as to distinguish between the
'deserving' and 'undeserving' poor was recognised by the tenants. One
woman commented:

See, years ago, it was a struggle to move onto this estate, they were
selective, they had a bit of ... without actually being told, there was
discrimination.

(Mrs Goody)

Another pointed out exactly what the implications of this discrimination
were for ethnic minorities and for 'problem families'.

We were told that after the estate had been occupied for twenty
years they could put anyone in, know what I mean? and, er, how can
I put it, I don't want to be anti-racial do I? You know like black
people, Greek people. Up until then I think they were very selective
with their tenants, you know I mean very, ...careful. But, even so
after twenty years ... we have one or two problem families, but it
was explained to me that they do that, hoping that the people that
they put them in with will be good for their way of living and it has
worked out.

(Mrs Taylor)

Although the respondent talks about ethnic minorities and problem families in the same paragraph she did not mean to imply that all ethnic minority families were problems. However, it is clear from the statement that for a long time both ethnic minority families and households which the authority might regard as a problem would not be allowed on the estate.

Although six women commented on the selection procedures, none of them criticised them. One positively approved and had no doubts about her place in it:

> Because I would say, without being, you know, snooty, the tenants were selected. Like getting into a school, there is a selection isn't there? And we thought it was good, yes, good.
>
> (Mrs Quick)

Another was less certain of her position and stated simply:

> I think I was very lucky to get this place.
>
> (Mrs Leary)

It is significant that this last speaker was Irish, married to an Irishman; she was a mother of six children and had a full-time job as a cleaner, he was an unskilled labourer in the building trades. The interviewee who had no doubt of her place in the selection procedure had two children, worked part-time in an office and was married to a policeman. The LCC selection procedure seems to have reinforced both class-status and racial divisions. This is in accordance with other studies, such as Gallagher's (1982), which found that tenants who defined themselves as respectable and who were thus defined by housing management enjoyed their status and their physical separation from other less 'respectable' tenants on other estates.

One further aspect of housing management which contributed to its popularity was the high standard to which it was kept for the first ten years after it was built. The external landscaping and the internal common areas were preserved by the meticulous attentions of the caretaker. He retired approximately ten years after the estate was opened. His passing was much regretted:

> But when we came here it was quite nice, we had a nice caretaker and everything. He kept the estate lovely but now it's going to rack and ruin.
>
> (Mrs Webster)

The effort which the caretaker brought to his job gave the estate an indefinable extra quality, to one woman at least:

But it's still a lovely estate if it was looked after properly. Well I'm not saying it isn't looked after, but there's something there that isn't as it used to be.

(Mrs Leary)

The pride which the caretaker took in his job was shown publicly in the way in which he was seen to be carrying out his work. One tenant reminisced:

Eight o'clock in the morning, the porter would be down there sweeping the path, sweeping the lift, the caretaker would be up seven o'clock and you would see him walking along, if he saw a piece of paper on the grass he would pick it up.

(Mrs Goody)

He wore a uniform and would be out every morning, supervising his men. The caretaker was so bitter about the poor way in which the caretaking duties were carried out after he retired that he refused to be interviewed.

As an ex-Services man the caretaker supervised the ground staff or porters with exact discipline. He adopted the same attitude towards children playing on the grass. The estate had been designed with many grassy, open spaces. Whilst it is clear that the object of the design was to bring greenery onto an urban housing estate, it is not clear how these spaces were intended to be used. The caretaker, it seems, was convinced that the grass was intended for purely visual purposes. He kept the grass well and did not allow any children to play on it.

Course the original caretaker was quite strict, he used to come round with his little dog and chase the children off the grass. And the dog got to know what his duty was and the dog would chase them off as well! It was rather funny 'cos the dog could run faster than the caretaker.

(Mr Nash)

It was not only children who were warned off the grass, adults were prohibited from walking on it as well.

You're not really allowed on the grass. When we first moved onto this estate we had a marvellous caretaker. Mind you, at the time we always used to think, ooh, he was so strict! He was an ex-naval man and this estate was kept beautifully. Nobody was allowed to walk on the grass *at all* and he kept it like that, you know.

(Mrs Taylor)

Whilst the caretaker's attitude was a mixed blessing, as the comment by Mrs Taylor shows, his activities helped to raise the appearance and status of the estate. Obviously the landscaping, layout and design helped: if the original trees had not been kept and more planted it is doubtful whether the estate would have been so attractive. Nevertheless, as Mrs Ivel pointed out, on the Crescent it was possible to maintain the appearance of being on a private estate since none of the usual problems associated with council estates were in evidence:

> This estate was always a credit to him [the caretaker] in as much as
> it was nice and clean and tidy – the kids were a bit of a problem
> because it was a younger estate then but they never got out of hand,
> we never had vandalism or graffiti or broken windows, nothing like
> that, it's never been like that.

> (Mrs Ivel)

'Just ordinary hard-working people'

The decency of the Crescent was not only seen to lie in its location, design and management but in its tenants and their attitudes. Comments were made on the lack of quarrelling and noise and on the good behaviour of children, as this conversation between tenants demonstrates:

> Mrs Dennis: Most people that come here say, what a decent, what a
> decent estate it is, they haven't seen it as we saw it years ago, but it
> still is as far as some places go.
> Mrs Sugden: ...they behaved themselves.
> Mrs Dennis: Mm, that's right.
> Mr Sugden: We don't get any real aggravation.
> Mrs Dennis: No, you never hear quarrelling among tenants much, or
> nothing.

Six women commented on the 'niceness' of the other tenants on the estate. For one of these, the quality of her neighbours made up for the fact that she had had to move onto a council estate at all. Miss Read and her brother had owned a chemist's shop in Shoreditch which had been compulsorily purchased by the LCC, much against their will.

For two households, moving on to the Crescent had meant that they were able to move away from neighbours whose behaviour they saw as being rough and difficult to live with. One woman had been living in Hoxton, in a council maisonette with her second husband and three daughters. Her neighbours there were fond of getting drunk and would return from the pub and urinate over the balcony. She was anxious to escape from that environment, for her own as well as her daughter's sake.

For another woman, her reasons for preferring the inhabitants of the Crescent Estate were entirely the reverse. Her previous experience was of living on an estate run by a Housing Association: there most of the tenants were in households headed by men with safe, permanent jobs. Although her husband earned a good living as a parquet-floor layer, he was not in this category. She felt that the woman next door, who was married to a retired detective, had ignored her for that reason. However, her other neighbours had been friendly.

Mrs Taylor's opinion of other people on the Crescent Estate was echoed generally.

They were all very nice people here, keeping the place nice.

(Miss Young)

The majority of the interviewees' attitude towards the LCC tenancy agreement confirmed an impression of them as decent, law-abiding people. Council house tenancy rules have been criticised by authors such as Gallagher (1982) as being 'oppressive and regulatory'. White, however, in a historical study, demonstrated that rather than finding the rules repressive, the majority of tenants accepted them because they only codified what the tenants would have done anyway (White 1981).

This last attitude was true of the respondents on the Crescent Estate. They were mostly undisturbed by the rules. Mrs Webster was dismissive,

Oh, I didn't take no notice of them, all that you couldn't have no cats, we didn't have no cats so that didn't bother us.

Six interviewees thought that the rules represented common sense practices. Mrs Leary elaborated on this with respect to cleanliness and tidiness.

Well, there was nothing wrong with the rules, everything was quite alright, if you kept to them, there was nothing too strict, just no shaking mats over balconies and all that lark was on your rent book, but I mean, who should shake carpets over their balcony, there's people underneath going to get it aren't there? No, and to keep the chute clean and to keep the staircase clean – that's all their rules and, well, you mustn't destroy their property must you? Apart from that they weren't too strict, their rules weren't too strict at all, they were just normal.

Two of the interviewees did break the rules by keeping cats, when they first moved onto the estate. One of them was worried about it and took elaborate precautions to prevent himself from being found out. He used to keep the larder empty and if there was a knock at the door, then the cats would be shut in there quickly. The household also used to have lace curtains up to prevent the cats from being seen. This man described the rules as 'pernickety' and was, in this sense, an exception. Another woman told me of a resident who took in homework and the noise of her machine could be heard from her flat; this was also forbidden. However, it seems to have been an isolated incident. The most common way in which the rules were broken seems to have been by people making minor unauthorised alterations to the flats, such as putting up TV ariels and taking down picture rails. These incidents were reported by hearsay rather than by the interviewees themselves.

Apart from these occasional transgressions, most tenants seem to have kept most of the rules most of the time. For example, in response to the question why tenants who had drying cabinets in their flats did not use the laundry, Mrs Goody replied:

> Well, people just knew they couldn't use them, if they weren't in those flats. That is how people were when they came, they were nice people.

Within the tenants' perceptions of themselves as decent, well-behaved people there were differences and divisions. Mr Owen reported,

> ...there was one or two that put on airs and graces which you'd find out afterwards was only affected. I'd say we were all more or less from the same sort of level I would say. Just ordinary hard-working people.

Other respondents gave more importance to these differences than did Mr Owen: these divisions will form the subject of the rest of this chapter.

Airs and graces

The phenomenon of different sections of working-class people differentiating between each other in terms of gentility and status has been well documented. Stacey, in her study of Banbury, found that three clearly discernible groups could be identified, 'rough', 'ordinary' and 'respectable' (Stacey 1970).

The respectable males were engaged mainly in skilled manual work, although some did clerical and some semi-skilled manual work. The ordinary had mainly semi-skilled manual occupations, although some were engaged in skilled manual work and some unskilled manual work. The rough generally tended to do unskilled manual and semi-skilled manual work.

However, Stacey found that it was the way in which women kept their homes which contributed greatly to definitions of class-status. Thus, although her husband might do semi-skilled manual work, whether the household would be classified as rough, ordinary or respectable would depend entirely on the wife's efforts.

Stacey noted that among council house tenants the council houses and therefore households that had least status were those which were built for slum clearance. Dwellings built post-war had greatest status since there was a considerable social mix within them. Tucker, in a journalistic study also found that there could be status differences within estates, sometimes based on geographical location and sometimes purely on social difference (Tucker 1966).

In Stacey's definition the roughs were households where the breadwinner was not in regular employment, where the house and children were not kept clean and whose family members were in trouble with the police from time to time. None of the people interviewed on the Crescent fell into this category: indeed the selection procedure was quite clearly aimed at keeping them out. Distinctions between the 'respectable' and the 'ordinary' could be observed and these were largely discerned through the activities of women.

'Keeping the place nice'

Status divisions between tenants found their focus in homemaking and childcare. For the interviewees, these were women's responsibilities.

That women did and should take responsibility for housework and bringing up children was taken for granted by all the people interviewed.

Perhaps the most typical attitude was summarised by one man. When asked about his home in terms of doing housework, he replied:

Of course that I didn't know much about housework because it didn't have a lot to do with me. [Laugh] Just helped out now and again.

(Mr Owen)

The extent to which women were helped with housework varied. The key factor in how much help they received was whether they had waged employment or not. Mrs Webster, who did not, received no help at all from her husband and in fact described how she had to rush around on Saturday to get all her housework done before her husband arrived home from work. Mrs Quick who worked part-time received more help from her husband, who was a policeman. He vacuum cleaned and polished, while she washed, cooked and ironed. It was not an equal distribution of labour, however. Mrs Leary who had a full-time job and five children got her children to do housework. She managed by being highly organised.

Women, then, were the directors of the household. The standards to which they kept their houses and looked after their children were used as a matter of judgement by their peers.

To a certain extent standards of childcare had already been set by the caretaker and the porters. The caretaker's activities in chasing children off the grass have already been commented on. An informal policing system also worked since there were about six porters as well as Mr Murray, the caretaker. If children were seen doing something wrong then either a porter or Mr Murray would mention it to the parents. Mr Murray also would tell off the older boys if they were making a nuisance and apparently commanded some respect from them. Some burden of controlling children was thus shouldered by the estate management team.

On the other hand, the caretaker's and the porters' attitudes reinforced a standard of respectability. To be a good mother, women could not let children roam where they wished. Children had to be prevented from running on the grass, playing on the staircases and in the corridors of the flats: precisely the places where it would have been easy to supervise them. The respectable appearance of the estate meant that mothers had to enforce high levels of control over their children in order to preserve that same appearance and atmosphere.

Comments were made by interviewees about 'others' who failed to live up to their standards. Mrs Dennis said of her and her friend's children:

They never stood up there and made a noise like some of them do.

Mrs Goody used the criterion of childcare to define the Crescent's superiority:

This is one of the best estates in London, I don't know if you've been to any other estate, this is the best estate in London, we did not have children running in and out of the laundry.

To add a further level of control, mothers who shouted loudly at their children were also criticised:

...the ones that did not come from Lowfield, well, they stuck out like a sore thumb, because you, you know, used to hear them shrieking over the top, you know, calling the children in.

Some women coped by sending their children over to the park to play. There were problems, however, with a busy main road to cross and: 'You get these men that are not very nice, even adults, teenagers.' There was one play space on the estate, but it was limited. It was also in the centre of the estate so that noise disturbed the surrounding residents. Whilst women who had older boys, or children of varied ages who could look after each other, could manage, others found it more difficult.

One woman in particular, found bringing up children in flats to these standards of respectability an enormous strain. Not only did Mrs Ivel not want her children to play where it was forbidden and be free from harassment, she was also concerned about who they played with, that is whether the other children would be respectable enough. She described her anguish graphically:

> Now I used to make a point of rushing around to do all my work in the morning, then in the afternoon I'd take them through to the shops, or we were fortunate in having a park or else I'd take them to the park to play. So that means that it puts a terrific strain on the woman that she's got to get all her work done in the morning to get her kids out of the flat in the afternoon you see. Well, once they start school they want to mix with their peers you see, but I thought well, I don't really like them going out to play so I used to have kids in to play in my place because I'd rather be looking after other people's children and know that my own were alright than sending my own out to play but it came to a point where you just had to let your own children go out to play and it used to worry me, it used to worry me sick, so if I let them out to play, once every quarter of an hour and I'm on the third floor, I'd be rushing down to look at them to see if they're alright.... So you were under a terrific amount of stress and strain the whole time. There's no doubt about that. No doubt about it.

Whilst many surveys have described the problems of bringing children up in flats, this comment of Mrs Ivel's, which has been quoted at length, illustrates the complexity of the problem. It was not just a question of where children would play and their physical danger. There were the problems of abuse by adults and also those of elitism or snobbery – who the children would be playing with, whether they would learn rough manners or speech from them.

Doing the laundry provided another point of social differentiation. The issue was over the hanging out of washing. This was regarded as such an unacceptable activity that LCC rules forbade it for flats.

The matter had been raised in full Council when a woman member asked the Chair of the Housing Committee whether tenants on three of the new mixed-development post-war estates had laundry facilities and if they had, why they were not being used. She went on to ask whether tenants could be asked to use the laundries opposite so that adjoining owners would not complain that washing was being put out to dry. The response was that the tenants did have laundries and did use them and that:

> Tenants are constantly asked not to display washing on balconies but in spite of this some clothing is hung on balconies to dry or air. Every effort is made to prevent it.[4]

Presumably members of the LCC felt that hanging out washing lowered the tone and spoilt the appearance of their 'best' estates.

Some members of the Architect's Department disagreed with this attitude. Cleeve Barr, who headed the Research and Development division of the department, wrote in his book *Public Authority Housing* that he thought that a south-facing balcony was a very sensible place for clothes-drying. He made a comparison with the Continent where picturesque views of clothes-covered balconies were much admired by English observers (Cleeve Barr 1958). The architect who had designed the Crescent Estate agreed with the mainstream of official opinion and thought that washing hung out on balconies would spoil the appearance of the estate.[5]

Some of the women tenants also agreed with the rule. Two tenants discussing it talked about the rule and one said:

> We wouldn't put it up high, would we? Like one or two people I notice do now, but we wouldn't. It looks slummy doesn't it?

One woman did disagree with the rule. Mrs Webster, who described herself as 'rough and ready' did hang her washing out.

> We got a cupboard in the kitchen where you can dry, but I've never used it, never. Nobody does. Turns your washing yellow. And look at the gas it takes. On sunny days we have them out here, on the balcony, but when we first moved here you wasn't supposed to do it, you know. 'Cos we were here for six months and I'd already got my washing hung on a Tuesday, and hung out, and they fetched three inspectors round. One of them turned round, he said, 'That's not the way to dry your washing, you've got them cupboards.'

However, Mrs Webster, in keeping with her overall sense of respectability, had her reservations about how far she would go in breaking the rule:

> But the only thing I don't like is hanging it out Sundays. I don't think they should hang it out Sundays, but they do. Most of them do

so. Well, I never used to years ago, but they do everything now, don't they?

Washing for a growing family and drying clothes in the absence of a washing machine and adequate drying facilities must have posed problems for women tenants. However, part of the hall-mark of respectability was being able to manage, despite difficulties. Mrs Dennis struck a note of tough defensiveness when asked about bringing up a family of six boys in a three-bedroomed flat with no outdoor hanging space or access to a laundry:

> There was eight of us, when I had my children at home and I've always done my washing haven't I? I've had no problems.

Women's responsibilities for home making extended as other authors have argued, into household strategies for increasing the household income. In Chapter five it was argued that the LCC provided the shell of a dwelling, which, in order for it to become a home had to be carpeted, curtained, furnished and heated. Cookers, fridges, vacuum cleaners and other necessities of life had to be purchased.

During the 1950s married women entered the labour market in increasing numbers. Surveys undertaken during this period showed that the increased income which women gained was used to buy necessities and goods for the home (Tilly & Scott 1978). The women interviewees on the estate were typical of that period in that few worked outside the home when their children were very young: however, by the time they moved on to the estate some had school-age children or teenagers and these tended to be the women who worked outside the home, full-time or part-time.

Mrs Goody described how people were able to build up nice homes which, she said were kept immaculately – so that as the cliché goes, 'you could eat your food off of the floor'. Mr Owen described how his wife had been immensely house-proud and had spent hours of her time polishing the vinyl floor tiles throughout their house.

There were inevitably status differences which arose between households who could afford more expensive furnishings and those who could not. Two women said that they did not invite people into their homes for that reason, because they did not want gossip about what they did and did not have.

Not only did this kind of elitism encourage isolation but the juggling act of combining housework and waged work and in some cases childcare in itself caused stress and strain. Again there was a denial of the importance of this. Mrs Quick's comment: 'You're not married? Think well.' was suggestive of the strain she felt in her household responsibilities.

Other women experienced more practical problems associated with waged work. Mrs Harris, for example, had to resolve the demands of two part-time jobs and housework. Although she denied that this was a problem, her account of her routine shows the pressure she was under.

[Did you work part-time?]
No, full-time. Actually I had two jobs because I used to get up and do office cleaning in the morning, so I used to be up and out at five-thirty in the morning, and then I went straight on to (a civil engineer's) till about four o' clock.
[How about shopping?]
Well, I used to do that in my lunch hour, you know. Being at Park Lane I used to take a bus and go down to Edgware Road, along the, you know, Church Street Market, is it? Yes, ...because I wasn't tied to time, as soon as I was finished, I sometimes was back for one, other times I was back, getting my teas (Mrs Harris was a tea lady in this job), I used to do all my shopping and take it home.
That was no problem really.

Mrs Harris told me that she cooked a meal every night and her husband cooked a meal on Sundays. She did a little housework every night too.

...so that it didn't get on top of me. At that time, the flat was a decent size, so I had a washing machine and of course it was pretty easy going, but other than that I had no problems.

Mrs Harris also said that when she went to the pictures she had fallen asleep during this period. Given her daily schedule, this is not surprising.

For women their position was never secure. In order for their status in the world to be affirmed their homes had to be made and re-made every day. This required a constant effort of assessing and juggling conflicting priorities. It was not just a question of being an adequate mother or an adequate housewife. Mrs Ivel, for example, had aims higher than that. Her concern was with respectability, with making sure that her children were playing with the right sort of other children and that her flat looked as good as new. As a consequence the contradictions between housewifery and motherhood were heightened, yet at the same time required resolution.

Men's responsibilities

Although homes were made for men, men were expected to make a major financial contribution towards their upkeep. The interviews only provided limited evidence from men: however they did confirm expectations that men worked long hours to cope with high rents and increased domestic expenditure. At the point in the family cycle when children were young, men might be expected to work overtime hours to deal with the extra expense of a financially dependent wife. The two men interviewed did provide some evidence for this double circumstance.

> ...I used to work long hours, and see, the funny thing is, I assumed that like when I got married, that in a way your life carried on much the same. You go out and enjoy yourself, but it isn't that way, obviously your money's got to be channelled into different areas, especially when there's children, it really puts the kybosh on it. So I sort of really had my working life, in those early days I was on night work, I used to work seven nights a week, really it was work and bed.
>
> (Mr Nash)

Men's lives were simpler than women's in that there was no overt conflict between their family lives and their jobs: the one existed to support the other. Although in an ideal world men who worked long hours could be accused of neglecting their children, in the 1950s this would have been considered absurd, since it was the mother–child relationship which was considered paramount.

Concluding comments

Interviews with tenants who moved on to the Crescent Estate when it first opened bear out the strength of Stacey's observation that 'in social class home and working life meet' (Stacey 1970:148). Not only is the man's occupation important, but how his wife uses that money to create a home, how she runs it and how acceptable both of them are to friends, colleagues and neighbours is crucial in determining status.

This chapter has attempted to show the stress which such class-status divisions place upon women. Whereas homes might be said to be created for men (although ironically the effort they expend in earning money might drastically reduce the time husbands spend in them), women have to live for their homes. Not only do the conflicting and irreconcilable activities of waged work, childcare and housework have to be juggled but the home has to be made and remade afresh everyday. The amount of autonomy which housework actually gives is small but it is sufficient to determine class-status to a significant extent.

It is remarkable how little attention has been paid these differences, by architects and social scientists. As we have seen, the architect for the Crescent Estate thought that he was providing rented housing 'for all', although he did admit that he had a middle-class view of it. The Vice

Chair of the Housing Committee wanted to provide rented housing kept so well that 'even the Queen could live there'. There was an ambiguity about the provision – the grassed areas looked beautiful when they were maintained but were unsuitable to be used for children's play, the stairs similarly were not designed for unsupervised children's play. The architects designed with the imagery of middle- or upper-class homes, wanting everyone to have the privileges which they had enjoyed. However, being male, they failed to appreciate the importance of the privileges which middle-class women had enjoyed before the war, such as childcare in the form of nannies and nurses and paid help with the housework. Working-class women then had to battle against the odds, aiming for the appearance of middle-class life styles on working-class incomes.

Feminists similarly have paid little attention to the class implications of home-making. A political stress on the common experience of women and a justified focus on inequalities between men and women have led to a consideration of housework as unpaid work. There are clearly differences of interest between women and men in home making with women having a greater portion of the work and responsibility and a lesser portion of the leisure and status in a majority of households. Yet there is often a joint purpose in home making that goes beyond the desire to have a family. This is the desire to have and maintain status within the world.

The surprising aspect of these interviews for me was the pride which women had in their homes. I had expectèd to find anger and resentment expressed – anger at lost opportunities for career or travel and resentment at the unfairness of having to take responsibility for domestic work and childcare. Instead I found a resigned matter-of-fact approach to domestic work summed up by Mrs Taylor who said:

> I mean the woman's got the job of bringing up the family haven't they?

Only one woman gave a hint of digruntlement with her responsibilities in her advice, in a wry aside, to think hard about getting married.

Whilst the preceding chapters have emphasised the structural inequalities between women and men in housing provision, the interviews do not show these inequalities as being keenly felt. The interpretation of interview material raises theoretical problems with regard to the interviewees 'real' motivations and intentions. Other authors, such as Hunt (1980), have sought to explain their respondents' apparent aquiescence with their lot as women by emphasising the social structures which have moulded their consciousness. Similarly, I have shown in the chapters on the location of housing and design the social and physical arrangements which make it seem utterly natural that these respondents should shoulder the main burdens of domestic responsibility within the home.

What none of the policy makers and architects considered is that working-class women and men would see themselves as moving 'upwards' through the social structure through having better housing conditions. Instead of becoming citizens in a classless society these respondents saw themselves as having bettered themselves and therefore stood apart from other working-class people. The strength of this assertion is illustrated by the fact that none of the interviewees mentioned other council estates nearby, except in a disparaging manner, in their discussion of the estate.

Whilst the wish to have a respectable home could on the one hand be portrayed as a bitter struggle against overwhelming odds by women, on the other hand, it could also be portrayed as a mean-minded undertaking, fuelling vanity, insecurity and petty jealousy. Certainly respectability may lead to isolation. This was evident in the interviews described and Stacey noted that the more respectable the working-class family, the more they kept themselves to themselves as regards neighbours. Furthermore, the desire to be superior undermines solidarity and community between women as differences are magnified. One of the distinguishing features of a 'rough' working-class family in the 1950s was the extent to which they were dependent on neighbours for borrowing, baby-sitting and other small favours.

What is also striking about these interviews is how successful the architects were in making popular an unpopular form of dwelling. Whilst it is true that it is likely that first tenants who had lived on an estate for twenty-five years would like it for some reason, the enthusiasm with which the interviewees discussed the external appearance of their estate was unequivocal.

The architectural form of the estate was unlike a private development, in its mixture of flats, houses and maisonettes. Furthermore, the flats had features which are nowadays considered to be detrimental to good management – open-access galleries, no private gardens, no 'defensible space'. Yet the interviewees made comparisons with private estates and two, after the interviews were over, seriously discussed the possibility of exercising their 'right to buy' under the then 1980 Housing Act. In the previous chapter I argued that the LCC architects had used the imagery of an aristocratic landscaped park to create an aesthetic language for their estates. These first tenants' response to that language suggests that actually the overall atmosphere of a design and the meanings which tenants attach to it outweighs individual elements of construction. This emotional–spiritual aspect of housing is difficult to quantify, but as these interviews illustrate, is significant in determining people's happiness.

Similarly, attempts at understanding gender and class have been limited by a rationalist approach to home life. Housework has been seen as work and housing not considered at all. Only Davidoff and Hall's study of nineteenth-century bourgeois families has considered class-status, gender and home life in any depth, from a feminist perspective (Davidoff & Hall 1987). Yet the interrelatedness of gender divisions and class-status was marked in the 1950s, as has been shown.

Chapter eight

From there to here

In the last five chapters I have shown in detail how, in the period following the Second World War housing was provided in a way which supported the belief that married working-class women's main responsibilities lay in looking after their homes and families. This belief also structured ideas about the location of new housing, which laid the way open for married women to be exploited as low-paid workers. An emphasis on the centrality of the nuclear family also disadvantaged single women, divorcees and widows in terms of housing provision.

Within these broad generalisations there were contradictions between the intentions of policy makers and the circumstances within which they were working. Thus attempts to decentralise London fully were thwarted by pressures from industrialists and the existing structures of land values among other things. There were also differences between the intentions of architects and the way in which tenants perceived and used the estates which were designed in the period. In this way architecture does not 'reflect' in a simplistic manner the attitudes and customs of society – rather it is formed within it, is part of it and on occasion lasts beyond it.

In this period following the Second World War the domestic lives of some working-class women and middle-class women merged in circumstances and surroundings as domestic service declined to an absolute minimum and housing standards rose. This tendency continued in the following four decades. The division of housing form between flats in the city centre and houses in the suburbs also continued to be a dominant theme. The continuity of these tendencies and themes will be examined in the following pages before I draw out some conclusions.

Housing form

In Chapter three of this book the ironies and paradoxes of flat building in London and other urban areas after the Second World War were explored. Although central government had espoused a family policy and despite evidence that flats were an unsuitable form of dwelling, flats were built because of pressures on land and a desire to have an urban environment with public open space.

This trend towards the building of flats for rehousing families in sub-standard accommodation continued throughout the 1960s. Dunleavy (1981) has argued that local authorities in inner urban areas were particularly constrained because suburban authorities closed ranks and were unwilling to allow any of their land to be used to house overspill populations. Electoral considerations played a part in this process.

Partly because of a change in subsidy from the Ministry of Housing and partly through pressure from the construction industry and architectural profession, many estates with high flats were built during the 1960s. Although these have tended to dominate the imagery of local authority housing, it is important to remember that even in 1981, before the current wave of council house sales, flats of over six storeys constituted approximately only 7% of all local authority stock.

Architects had been excited by the possibility of high-rise building since the early 1920s. In Chapter six I explained the vision of living which the modernists aspired to: during the 1960s the ideas were still current and became part of an established orthodoxy. The great architect of the modern movement Le Corbusier had neither designed a tower block for family housing nor advocated tower blocks as an ideal housing form. Indeed, in 1957 he finished a housing block in Marseilles which took a different form from a tower block. The Unite d'Habitation was a 'slab' block which not only contained flats, but had a crèche and shops.

The model for tower blocks came from Sweden where tower blocks had been built successfully after the war. For architects the issue in high-rise housing in the 1960s was not whether high-rise was a suitable way of providing housing, but whether tower blocks should be built, or slab blocks or deck access schemes. This preoccupation was illustrated in a survey carried out by central government in which tenants (housewives) were asked if they preferred to live in slab blocks or tower blocks. The response was that no one form of dwelling was preferred over another, but what the women interviewed did dislike was looking out onto concrete. A strongly favourable response was given to looking out over greenery (DoE 1972).

High-rise housing drew important themes of modernism – standardisation of building components, a use of the most up-to-date in services technology and a dramatic break with the past. It is difficult to assess what ideas about women were connected with high-rise because much of the high-rise programme was contractor-led and derivative of architectural ideas rather than being part of one well-known architect's work.

It is important to remember that architects in the 1960s were still influenced by ideas from the 1920s and 1930s, which saw the introduction of technology as emancipatory in itself. Certainly the possibility that women would be 'free' to take up waged work was raised by modernising the home, in particular through the introduction of central heating. It was assumed that through living in highly serviced, compact but 'efficient' flats, women would be freed from the drudgery of household chores and would be able to pursue waged work and child rearing more readily. These ideas were reversed by sociological studies of mothers of young children in high-rise flats which showed that they were more prone to depression and ill-health (Gittus 1976).

Whilst inner-city local authorities were building futuristic flats, rural authorities and speculative house builders were constructing estates of two-storey terraced and semi-detached houses. More women were being drawn into the waged labour market: this was recognised in the Parker Morris Report which was first published in 1961 (MoHLG 1961). However, the accommodation to women's participation in the formal economy which was made in design terms was not that the physical shape of houses would change radically, but that more appliances would be purchased by households and fitted in their homes.

Far more radical solutions were available to housing authorities. These ranged from household appliances being supplied by them with the dwelling to the provision of nurseries and play schemes on housing estates along with other communal facilities such as restaurants and laundries.

Ironically, in an era when more married women were physically leaving their homes for employment, house design changed to place a greater emphasis on family life as a whole. In the 1960s architects and builders experimented with 'open plan' arrangements whereby there would be a free flow of space between the kitchen–dining areas and the living room. The architectural aesthetic reasoning behind this was a legacy of Modernism – the notion of opening up space and allowing a free flow, so that a house interior could be experienced like a Cubist sculpture. This arrangement may have been more acceptable than hitherto since by this time there could be no pretence that any household other than the very rich had servants. The open house plans of the 1960s lost favour in the 1970s and 1980s.

Although it not clear why speculative builders returned to more traditional arrangements, local authorities and housing associations reported that tenants preferred separate kitchen–dining rooms to avoid food smells permeating through the house and the sight of unwashed dishes.

An 'open-plan' arrangement led to women in particular having less privacy within the dwelling as even their work room, the kitchen, became part of the 'family' space. Possibly the shift from open plan in the 1970s might have been a reaction to this, as people demanded individual spaces over which they had control.

Thus, although profound changes occurred within married women's relationship to their families in terms of working outside the home in this post-war period, surprisingly little changed in the physical shape of the overwhelming majority of new-built houses. Only in council housing were new forms attempted and here architectural concerns were with urbanism rather than the changing role of women. For all the modernists' discussion of new ways of living, the one major social change which occurred in the post-war years passed unnoticed by them until the Women's Movement emerged.

Housing and families

Further changes were occurring in the composition of households. A trend towards the formation of smaller households had been apparent since the First World War. This trend continued after the Second World War, such that, as is commonly known now, the number of households living in a nuclear family arrangement has declined to a third of the whole (Central Statistical Office 1989).

The reasons for this trend were not that marriage was declining as an institution, but that simultaneously, a greater proportion of the population lived to an older age and the divorce rate increased so that now approximately one-third of marriages end in divorce. Thus the rate of household formation and dissolution is greater than a conventional model might suggest. Instead of a couple meeting in their early twenties, marrying and having a family when they are twenty and thirty years old, then living as a couple alone for ten or fifteen years before death, divorces are occurring while children are young and each member of the couple might re-marry and form another household. Women also tend to outlive men, so that rather than a couple growing old together and dying within months of one another, the husband tends to die first leaving the wife living on her own.

Housing policy during the 1960s and 1970s was mainly premised on the conventional model. Owner occupation was supported by both Conservative and Labour governments. Although owner occupation in itself need not necessarily favour any particular household form, the internal rules of building societies made it difficult during this period for women heads of households to obtain mortgages. Moreover, the price of housing normally varies such that prices fall within the range of three to four times the average male income. Since women's wages during this period were approximately half to two-thirds of the average male wage (Reid & Wormald 1982), this meant that the obstacles to entering owner occupation were high for female-headed households.

Not only is this policy discriminatory in terms of gender, it is also discriminatory in terms of age. Young people who earn less find it commensurately difficult to finance housing costs – similarly people who may be only ten years from retirement age also have problems raising large loans.

Local authorities continued to pursue family policies in their allocation. The pressure for dwellings was such that it was impossible to cater for single people under retirement age. Points systems tended to be geared towards numbers of children and over-crowding, which again effectively excluded childless couples and non-familial households.

Although provision was made in the latter part of the 1950s and the 1970s to build housing for special needs, that is for the elderly and disabled people, the amount that was built was not high. Families of three, four and five people were the main recipients of the slum-clearance programmes and general needs programmes.

The 'permissive' sixties did bring some changes, albeit a decade later. Immediately after the Second World War unmarried mothers were regarded as either unfortunate or immoral. Local authorities provided special homes with strict regimes for those mothers who decided not to give their children for adoption. Women who had been deserted or divorced also bore a stigma. Attitudes changed and the Finer Report published in 1974 highlighted the plight of single-parent families. The 1977 Homeless Persons Act finally gave local authorities a duty to house single-parent families.

Whilst the Parker Morris Report mentioned hopefully that men were giving women more help with housework (MoHLG 1961), this was not based on any facts. Young and Wilmott, in their book *The Symmetrical Family* published in 1973, also suggested that men were becoming more home centred and were therefore taking a greater share in housework. These misconceptions were corrected when time budget surveys showed that men were not in fact taking a significant share of domestic work (Vanek 1974).

Some local authorities cut back on their maintenance of their housing schemes. Obviously financial constraints played a major part in this. The Greater London Council severely disinvested in public housing when it withdrew resident caretakers from estates in the 1960s and replaced them with 'mobile' caretakers. Thus, not only did the number of caretakers decline, but the informal policing which they undertook became an impossibility. The GLC, as a large housing authority, set an example to similar authorities to make cuts.

Housing design

In Chapter six the attitude of progressive architects within the LCC was shown as one of designing for a classless society – moreover a society in which working-class people would be able to enjoy the same advantages that middle-class people had hitherto enjoyed. In the late 1950s and 1960s there was a shift, on the part of some architects, to designing not for a classless society but for specifically working-class communities.

The inspiration for this came primarily from the publication of Young and Wilmott's book, *Family and Kinship in East London*. This study, which was first published in 1957, emphasised the importance of kinship ties and especially the mother–daughter relationship amongst the sample studied. Decentralisation was criticised as leading to a break-up of kinship ties and isolation of young mothers. Young and Wilmott's study was criticised at the time and more recently by feminists and others. Stacey, for example, pointed out that only 60% of the sample showed these strong mother–daughter ties. Furthermore, she emphasised that mothers could provide an important buttress against misfortune for their daughters, for example, if a husband either died or deserted them (Stacey 1970).

Nevertheless, a recognition that there could be an authentic working-class culture had an important influence on design theory. Two influential architects, Alison and Peter Smithson hoped to recreate working-class communities in 'streets-in-the-air' within housing schemes composed of flats and other facilities. The Smithsons thought that working-class streets were full of life and they wrote appreciatively of cars being repaired, people talking, children running about. They did not wish to preserve the untidiness of back yards and the by-law street: their idea was that the spirit of street life would be preserved, but in a new, orderly environment.

By building long flat blocks with elevated walkways onto which front doors opened, it was hoped to do two things – to separate pedestrians from traffic (cars would circulate at the base of the blocks and be parked there) and to recreate the working-class street at high level. On this 'street' it was hoped that children could play, women could stop and gossip and milk would be delivered by special electric carts. As the Smithsons put it: The flat block disappears and vertical living becomes a reality. The refuse chute takes the place of the village pump (Smithson & Smithson 1970:51).

Many thousands of housing units were constructed according to these principles. The most well known are Park Hill in Sheffield, Broadwater Farm in Haringey, Hulme in Manchester and the Mozart Estate in Westminster. The class to which they were addressed changed almost as the blocks were finished. Although no definitive studies have been carried out on individual estates it is possible to speculate on the changes which have occurred and in particular the way in which women's lives might have changed.

Stacey, in her study of Banbury, found that the working-class households who were most dependent upon neighbours were also the least respectable. Within her classification, the 'roughs' borrowed a lot, the 'ordinaries' had 'give-and-take' relationships and the 'respectables' were withdrawn and stand-offish. Poverty was a major factor in deciding interdependence (Stacey 1970). Although the original catchment for these inner-city estates would have been areas of slum clearance, it is likely that the 'roughs' who were to be rehoused would have been offered existing council stock, for as was seen in Chapter seven, new council houses tended to be offered only to carefully selected tenants. Thus, it is likely that only the 'respectables' and 'ordinaries' would have been rehoused on these estates, that is those sections of the working class who had less to do with each other.

The major social change which took place in British families in the decades following the Second World War was the gradual drawing in of married women into the waged labour force. This must have had a marked effect on patterns of neighbouring and street life. In Stacey's study of Banbury she noted that where working-class wives went out to work they tended to have a second set of friends and less time to talk to neighbours. Their dealings were also no longer reciprocal since they could not return favours – for example, looking after a child after school or taking in a parcel. Thus in any working-class areas there must have been a lessening of informal contact between women as they had less time to talk and pass the time of day with each other.

Consequently, it is possible to speculate that the street life which had been observed before the war was no longer possible – women were at work, children were at school and old people were in separate households of their own. Architects remained concerned with neighbouring, however, well into the 1970s and devoted much time in devising layouts for estates which would encourage informal contacts between people. Cooper has suggested that this can be generalised as a problem with architectural design – that architects are caught in a historical trap. If they design well for the present, by the time the building is built or has been in use for five or ten years, social circumstances may well have changed (Cooper 1978).

This concern with neighbouring extended beyond inner-city council estates. Many suburban estates, some owner occupied, were designed with carefully thought out cul-de-sacs and quaint alleyways in order to encourage human contact and achieve the nebulous goal of community.

In the mid-1960s the imagery of council housing changed again from futuristic concrete to homely brickwork. The mould-breaking scheme was Lillington Street, which was designed by the architects Darbourne and Darke. This was hailed as providing a 'middle-class' look to council housing – possibly because it was a reversion to brick from concrete and to emphasising the individual flat over a monolithic block (Boys 1989).

By the end of the 1970s, design intentions were in class terms broadly as they had been in the 1950s, to make local authority houses comparable in aesthetic terms to private schemes. The difference being that, in the 1950s, private meant private rented, whereas in the 1970s, it meant owner occupied. Thus local authority housing had to 'compete' with speculatively built housing by volume house builders. Since builders were only willing to attempt new forms of construction when it increased their profits, by reducing labour costs for example, dwelling design kept to tried-and-tested forms. Two-storey, single-family houses, terraced, detached or semi-detached, with gardens, were the norm. An attempt to create a new form for housing in terms of high-rise had failed.

Cleanliness and tidiness

In the previous chapters I have argued that women played a crucial part in maintaining class-status divisions. Through their standards of home-making and childcare distinctions could be drawn and judgements made. Ravetz has argued that the 'housewife' emerged during the 1950s – classless and replacing the middle-class lady and the domestic servant (Ravetz 1984). Whilst this was true in that domestic service had declined to almost nothing and middle-class women had to do their own housework, divisions were perhaps drawn in a new way. The divide lay between 'rough' and 'respectable', and 'respectable' could now include whole sections of women married to working-class men.

The growth of women's involvement in the waged labour force also contributed to this merging of status divisions. During the 1960s a widening range of household appliances became available to an increasing number of households, such as fridges and washing machines. During the 1970s other services became more widely available so that by 1986, 82% of households had a washing machine and 72% a deep freezer (Central Statistical Office 1989). These changes meant that homes could be made and kept to a high standard of comfort and cleanliness.

I do not wish to suggest that housework is no longer a problem, nor that for many women living in substandard accommodation keeping an acceptable level of cleanliness is not a battle. What I do wish to emphasise is that since the Second World War a wide range of households arranged along a spectrum of income now own appliances and fittings which make housework less burdensome. This means that the wife of a doctor might actually experience more similarities than differences in the conditions of her domestic work with the wife of a plumber. This was not the case before the Second World War.

Tenure

A major social change which has occurred in the decades following the Second World War is a shift to owner occupation as the predominating tenure. During the 1960s and 1970s local authorities had the right to sell their houses or flats. Owner occupation was also supported by both Labour and Conservative governments. Increasing numbers of working-class people became owner occupiers as owner occupation became a major means of redistributing wealth. Meanwhile the tenants who had been rehoused by local authorities immediately after the Second World War were aging. Social security payments were re-organised so that unemployed tenants could claim their rent from benefit. The passing of the Homeless Persons Act in 1977 meant that local authorities had a duty to rehouse households with dependent children who were homeless.

These changes meant that the more well-off sections of the working classes had an incentive to move into owner occupation. This move may have been a physical move, away from an estate which they saw as changing in social composition towards stigmatised groups such as single-parent families or unemployed people or it might simply have been a question of staying put and exercising the right to buy. This incentive was increased by the Housing Act 1980 and subsequent acts which gave massive discounts to council tenants buying their council houses. Between 1979 and mid-1987 over a million council houses were sold in England and Wales.

These changes are illustrated by the following statistics, drawn from Wilmott and Murie's book on social divisions and housing. They point out that between 1954 and 1984 the number of recipients of supplementary benefit doubled in the population as a whole. The proportion of recipients of supplementary benefit living in council housing increased sevenfold during that same period. By the early 1980s nearly two-thirds of single-parent families who were living on supplementary benefit were living in council housing. Perhaps the most telling statistics are those relating to income. Wilmott and Murie estimate that in 1968 the proportion of the third poorest households living in council housing roughly reflected the country as a whole, that is 31%. By 1983, 52% of the 30% of poorest households were in council housing. Conversely, whereas in 1968 46% of the richest 50% of households were in council housing, by 1983 this proportion had dropped to 25% (Wilmott & Murie 1988).

The racial composition of council housing has changed too so that now people from a number of different ethnic minorities are living in council houses. Wilmott and Murie argue that the most disadvantaged people are now concentrated in social housing. Council housing is in danger of becoming no longer general-needs housing for working-class people but of being welfare housing on the North American model instead. As such it

Figure 8.1   An unattractive council estate

is possible that the occupants of social housing, provided by local authorities and housing associations, are becoming increasingly marginalised from the remainder of civil society.

Coupled with this concern is also a fear of crime. Levels of reported crime have risen, particularly on council estates. Since many local authority tenants are women who are in a vulnerable situation through age or being a single parent this tends to stigmatise council housing further. Thus through these policy changes, the class composition of council housing has changed considerably.

The majority of households live in owner occupation. The divisions between those who live in comparatively reasonable housing – which is watertight, in an attractive environment and has full amenities – and those who live in substandard housing had widened. Only a minority of council housing is substandard housing (see Figure 8.1) and not all owner-occupied housing is in good condition. However, there is a sharp divide between owner occupiers who, by paying a mortgage, have accumulated a capital asset and those living in poor-quality council housing or privately rented housing with no capital assets, rising rent payments and a substandard environment.

Since 1976 the construction of new local authority houses has declined to a quarter of its previous level in 1985. The concentration of the poorest and most disadvantaged households in council housing has been noted.

Figure 8.2   Attractive social housing (Tardis Housing Co-operative)

This means that the gap between tenures has widened. There has been much debate as to whether there could be said to be distinct housing classes, with owner-occupied households in one and council tenants in another (Merrett 1982; Saunders 1989).

It seems that the argument that people in different tenures occupy different class positions is difficult to sustain. Firstly, it is not clear that council tenants and owner occupiers have different interests. There may be many similarities of income, occupation and attitude. Although council tenants may have no capital invested in their homes, some owner occupiers who have bought with 100% mortgages at a time when house prices are falling might have no more income and less security than them.

The fact that such a debate has taken place does illustrate the extent to which the position of council housing in class-status terms has been lowered since the Second World War. It has slipped, in England and Wales, almost to becoming a form of welfare housing. Aneurin Bevan's post-war ideal of providing local authority houses so good that no-one will want to live elsewhere has changed to re-housing homeless people. These changes have occurred partly because of fiscal arrangements rather than design changes; the class composition of tenants has altered. However, the ugliness of 1960s system-built council estates with their problems of dampness and condensation has also damaged the imagery of council housing. Ironically, the majority of new-build houses constructed by local authorities or housing associations in the 1970s and 1980s have been built to a higher standard (see Figure 8.2) than the bottom end of the speculative market in terms of attractiveness of design, space standards and finishes.

The preponderance of owner occupation as a method of organising housing has reinforced the pressure on households to spend a large proportion of their income on their housing and to keep their homes to a high standard of comfort and order. This pressure is material and keenly felt: some households would be homeless were they not to spend 35% of their disposable income on their housing alone.[1] Having made such a large investment, the logic is then to make it as comfortable as possible.

The British system of owner occupation, with its subsidies and high costs of housing reinforces women's dependency upon men. Since the passing of the Sex Discrimination Act finance organisations have had to give loans for house purchase to single women, or female-headed households. However, it is likely that women's weak position in the waged economy has limited their chances of getting housing in this way.

Whereas in the 1950s local authorities were keen to uphold certain standards of behaviour and were loathe to let their dwellings to say, one-parent families, there has been an increased liberalisation. Unfortunately in the early 1980s, just when local authorities were beginning to assess the implications of equal-opportunities policies, local authority housing suffered severe cutbacks. Thus some of the more sensitive policies which councils were beginning to evolve in terms of housing, for example, in offering accommodation to lesbian couples, or in coping with relationship breakdown, have had to be dropped because of the severe shortage of council housing.

It is not clear what effect this increase in owner occupation has had in design terms. Certainly volume house builders use architects to prepare layouts and to get planning permissions. Some innovatory schemes of flats have been designed in London's Docklands with swimming pools and other communal facilities. However, the drive for profit has meant that house building has been geared towards what people can afford and aspire to rather than need. Hence 'executive' bungalows are built in rural areas with twin garages, en suite bathrooms and expensive kitchens rather than complexes in the cities which would be ideal for low-income households.

Owner occupation, then, I would suggest has reinforced women's dependence on men, conservatism in design and high housing costs. This is not necessarily a function of tenure. Owner occupation in Britain is unique among European countries and is capable of being organised differently (Ball 1986). Having briefly described some of the continuities and discontinuities in housing over the last three decades, I now wish to make some concluding comments.

Concluding comments

House form and the reproduction of the labour force

In the early part of this book I showed how pro-natalist sentiments had shaped discussions of the re-planning of London after the Second World War. Concern was expressed that housing should be of a form that would encourage motherhood: leading proponents of the Garden City Movement, including the author of the County of London Plan argued that flats, where the population was 'cabin'd, cribb'd, confin'd' would act to reduce the birth rate. Pro-natalist ideas were taken up by a wide spectrum of political opinion after the war: nevertheless flats had to be built in inner cities because of pressures on land values and the resistance of employers to wholesale decentralisation.

It is difficult to assess the impact of these policies without further research. Certainly a 'baby boom' occurred in the decade following the Second World War. Since then the birth rate has declined and risen again. It would take a detailed study of differential birth rates in restricted geographical areas to determine whether flats really did discourage human reproduction.

Although Mumford's (1945) advocacy of motherhood over women's careers was so chauvinistic to be laughable it did contain a serious point. As I have shown studies undertaken before the war showed that flats were an unpopular dwelling form for families. Further studies of families in high flats elaborated how flats over five storeys in height were totally unsuitable for bringing up small children. This was not just a matter of preference but of physical and mental health (Gittus 1976). Yet in the 1960s many local authorities built high flats in city and town centres. Obviously there was a mis-match between the needs of children and mothers and housing form.

Furthermore, although one- or two-storey houses with gardens might seem to be the ideal dwelling form for family life, such housing does not answer all children's needs for play-space. A research bulletin published by the Department of the Environment in 1973 found that even on an estate with two-storey houses and gardens children's play was still a problem (DoE 1973). This was because apart from the children's own garden and their friends' gardens there was no other place to play: traffic was a problem as well as supervision.

The private market in owner occupation has reproduced the flats in centre of towns and houses in the suburbs scenario. With current house prices being approximately four times the average male wage, this means that for many young couples with small children it may be impossible either to afford a house rather than a flat or to buy a house in a metropolitan centre where both spouses have access to employment. The housing system at the moment, although based on a family policy, does not aid the bringing up of small children. In owner occupation, housing costs are highest when a family's capacity to pay is lowest, that is, when the parents are young and possibly earning a lower wage than in middle age. Thus the problems of housing form and simple physical reproduction remain.

What the changes in the housing system in the twentieth century have ensured is the orderly reproduction of the labour force. In the nineteenth century, working women moved freely through a number of different households and a variety of relationships. In Chapter two I contrasted this with the rectitude and orderliness of upper-middle-class households. As I have shown 'disorderly' family arrangements were excluded from access to council housing until the last quarter of the century. Thus unmarried mothers, lesbian couples, couples co-habiting, extended families were excluded from council housing unless they were rehoused under slum clearance.

Lodging has also been discouraged, through allocation policies, tenancy agreements, mortgage agreements and restrictive covenants. The only time when it was positively encouraged by either central or local government was during the Second World War when mobility of labour became a necessity.

A design result of this rigid conformity to a nuclear-family stereotype has been the monotony of much British family housing. Design briefs for both public and private sectors have been couched in terms of providing for a house for two parents and either one, two or three children. Although this is a stage through which many families pass, it is only a snapshot of the entire range of household formations. It is not surprising that such a rigid brief should produce such unimaginative designs. Respondents in Saunders' questionnaire on home life in three British towns referred to their homes as 'boxes' (Saunders 1989). It is as though the richness and painfully arbitrary nature of human relationships has been forced out leaving only an empty brick shell. Architects' attempts at variety and interest are only cosmetic applications to this bureaucratic desire for order, a need to ascertain ownership and apportion costs.

Gender divisions and housing provision

In the preceding chapters I have elaborated how family policies guided the relative locations of housing and employment in the period of state intervention after the Second World War. Men's position as prime wage earner was supported by relocating housing next to sources of employment for skilled men. Women's secondary position in the labour market was reinforced by creating pools of under-employed women in peripheral locations. Office employment was ignored by planners throughout the 1950s and early 1960s although offices were a major source of jobs for women.

Since 1979 state intervention in the location and financing of industry has declined dramatically, it being one of the policies of the present government to allow the free market to reign unhindered. Whilst unemployment has risen generally, women have not borne the brunt of it. Employment for women in part-time unskilled jobs has increased in the last decade.

Across the country problems associated with the 1930s have re-surfaced. There has been a building boom in the south-east with what seems like every available site being used for the construction of housing or offices. Roads have become much more congested, not only in London but on the motorways surrounding London and other metropolitan areas. Some towns in the south have expanded, such as Basingstoke and Swindon, as new firms providing financial services and industries dependent on new technology have moved there.

Meanwhile, the north of England and parts of Scotland and Wales have suffered a decline. Traditional industries have either declined or been decimated by lack of investment. Unemployment rates have risen disproportionately in comparison with the south-east.

These pressures have led to rising homelessness in the south-east and congestion on roads and public transport. At the moment, in 1991, we do not know how women have fared out of the boom years of 1984–1988. Perhaps employment opportunities have been increased for some groups of women and their lives made easier by greater access to car ownership. Or possibly others may be trapped in good-quality housing in semi-rural areas with restricted access to employment and facing the problems of a declining infrastructure of schools, transport and public health.

However, what does seem clear is that the issues of the location of housing, employment and transport are as live and vital as they were in the 1930s and 1950s. Even the Royal Family has been ready to make an initiative. Leon Krier's plan for Poundbury, a new quarter of Dorchester to be built by the Duchy of Cornwall, overturns the notion of zoning and hopes to attract a mix of housing, small workshops and shops so that car ownership will become unnecessary.

At a time when the necessity of town planning is being forced onto a reluctant government it does seem that the issues that feminist geographers have raised and which have been briefly examined in this book become more pertinent. Whose jobs are more important – men's or women's? Should not women's journeys to work be considered as well as men's? How can women be given equal employment opportunities? These issues are not solely concerned with the economy but with the location of housing, the provision of public transport, education and child-care facilities.

Housing and class

A theme to which I have returned throughout this book is that of order/disorder. I have argued that class-status divisions rested, in the 1950s at least, partly on women's responsibilities for housework, child-care and home-making. I also argued that during the 1950s the LCC architects based their imagery of housing estates on the aristocratic land-scaped park with its ordered views and pleasing prospects.

As I have indicated the moral pressure on women to keep a high stand-ard of housekeeping is strong. There is also a necessity for cleanliness in the care of young children, who are more prone to disease and minor ailments. It is not only vanity which inspires the wish for a clean comfort-able home: housework is easier to do with fitted carpets, central heating, washing machines and dishwashers. I make this last observation from personal experience, much resented! Household tasks now are less fati-guing than they were at the turn of the century when water had to be boiled on pans on the stove and coal fetched in from outdoors. The press-ure for cleanliness is still there, but less effort has to be expended to achieve it. The sexual division of domestic labour has not changed signi-ficantly despite married women's greater participation in the waged labour market. Hence it is not only that married or co-habiting women are undertaking more work but that they are still subject to social pressures to maintain high housekeeping standards.

The aspect of housekeeping which seems to be more crucial for coun-cil tenants now is that of the external spaces on estates. Since local authorities have withdrawn resources from caretaking many council es-tates have become befouled with litter, dirt and graffiti. I have argued elsewhere that this breakdown in municipal housekeeping has been mis-interpreted as a breakdown in law and order itself (Roberts 1988). Although little research has been carried out on caretaking, the experi-ence of the Priority Estates Programme suggests that the cleaning functions of caretaking are vital in an estate's improvement (Power 1987).

In the first six decades of the twentieth century, when renting was the majority tenure it was possible for class-status divisions to be expressed in terms of space standards and housekeeping. Now that owner occupa-tion has become the majority tenure, ownership has become a dividing line of status. Since subsidies have been withdrawn from council housing and kept constant to owner occupiers in the form of mortgage tax relief, disparities have become sharper between tenants living on the worst council estates and those able to afford comfortable houses. Again it should be emphasised that only a small percentage of council housing is in a degraded state and not all owner-occupied dwellings are in perfect condition.

Gender and class

Chapter seven revealed the depth of tenants' desires for a home which would answer their aspirations. These aspirations were not merely for a dwelling which was affordable, clean and decent – there was more than that. The manner of the tenants' response to their environment showed that they wanted homes which were in beautiful surroundings, which would be uplifting to live in. Moreover, they wanted to feel that they were living, if not in the best, but certainly in as good accommodation as people with similar or greater incomes.

The dark side of these aspirations was snobbery, narrow-mindedness and racism – all of which were expressed in the interviews, sometimes thinly veiled, sometimes not. I have chosen not to repeat tenants' comments which bordered on racism: nevertheless, some of the feeling that the estate had 'gone down' could be attributed to the elderly white tenants' perceptions of numbers of single-parent families on the estate. The good side of aspirations for a better life was the genuine appreciation of the natural environment – grass, trees and flowers. What I found particularly encouraging, as an architect, was the tenants' perception of the skill of the architect who had designed the Crescent.

It seems to me that the functionalist approach to housing which was adopted by some architects and policy makers alike after the Second World War has done tenants and people in general who still, after all have to look at council estates, a great disservice. People's needs in housing, women, men and children, cannot simply be expressed as numbers of dwellings, a shopping list of tasks or even as a simple formula 'houses-with-gardens'. On the evidence of these interviews, people have a greater sensitivity to their environment than they are often credited with. Because people are poor and have had to live in reduced circumstances does not mean that they dislike flowers, trees or a well-proportioned building.

These interviews showed the importance that home life had in terms of class-status in the 1950s. Whilst the concept of keeping pace with the neighbours is familiar, this is normally judged in terms of material acquisitions – washing machines, videos and other sorts of aids to living. What is not widely acknowledged is the extent to which women, as prime bearers of responsibility for the home, make and re-make that home every day. Furthermore, most working-class women do three jobs – housework, childcare and waged work. Women's wages go directly towards making a better home.

Whilst in the 1950s the majority of houses and flats were rented, a shift has occurred so that now the majority of houses and some flats are owner occupied. Successive Conservative governments realised the strength of people's desire not just for a shelter, but for a home that they could identify with in terms of status. Aneurin Bevan's hope of building council houses so good that no-one would want to live anywhere else was whittled away by spending cuts throughout the 1950s and 1960s so that the paramount ideal of politicians became building numbers, rather than making homes that people could be proud of. In the ideological battle over housing tenure, arguments over quality have become subsumed to arguments about tenure.

I have tried to show in this book how council housing, in common with owner occupation has been subject to practices which disadvantage women who live independently of men. Furthermore, although as I have shown, much emphasis is placed by both women and men on their home life, little practical support is given through physical design or through welfare measures, to the bringing up of children. Neither local-authority-provided rented housing, nor owner-occupied houses, need, of themselves, be disadvantageous to women. It is how these systems are operated, rather than the systems themselves, which are at fault.

Attfield, in her study of Harlow, commented that she thought that the 'pride and polish' with which women treated their homes in the 1950s could not be easily translated into the consumerism of the 1960s or even the 1980s (Attfield 1989). Whilst this might be true in cultural terms, in that the kind of furniture valued in the 1950s most definitely is not valued in the 1980s, nevertheless I think that the links between home-making, class-status and gender which we can see in the 1950s are still present in the 1980s.

Whilst studies of the home and interior decoration have been made, this aspect of cultural studies is relatively new. I would suggest that the home – its contents, location and women's experience of it should prove a more fruitful field for the exploration of gender and class difference than previous studies of the workplace.

The male domination of policy making and design has meant that the presence of a woman is indispensable to a home. While this remains the case it is difficult to imagine how a gender-free environment might be designed. However, there are a number of objectives, which, if achieved, could transform housing design.

The single most important change which could be made to the housing system which would benefit a large number of women would be to lower the cost of housing without reducing housing standards. Thus, married women, who earn less than men anyway, would not be trapped into unsatisfactory relationships with men for lack of an alternative. Lesbians and single women on low incomes would also benefit. 'Affordable' rents would be conceived in terms of women's wages, not men's. It is hard to see how such a radical change might be achieved without substantial public expenditure: however, this is not to say this could not be an aim of policy.

Architects have retreated after the aesthetic experiment of the 1960s and 1970s to designing much more traditional 'family housing', which loosely translated means designing houses with gardens in brick, tile and slate. I have argued that it is not the design which forces women to undertake the majority of domestic tasks. Yet, I have also shown how important design is to women's wellbeing and happiness. There are no clear routes for designers apart from understanding women's practical needs in housing and answering them in built form and engaging in the cultural values and meanings which women and men assign to their homes. This is no easy task and should not be restricted to feminist designers, who are few on the ground.

This book has shown the power of male domination and stereotyping in housing design. In it, I have considered the limiting effects these structures have on women. I have, in a sense, ignored men, since it seemed to me that women were subordinate. However, men are not absolute beneficiaries of gender division: they gain through not having to take responsibility for their domestic lives – but they also lose through having reduced contact with children and hence their own humanity. Such divisions cannot be life enhancing. I hope that feminism will not be a passing fashion and that the cause of feminism be placed firmly on the agenda, in architecture, housing and planning.

Living in a man-made world is not easy for women, whether they are heterosexual or lesbian, married, divorced or single, mothers, grandmothers or childless. Affordable, good-quality housing would benefit them all in their differing circumstances. Housing is too complex and too significant to be left to packaged programmes and simplistic stereotypes. A visually richer and more varied housing programme, which has been well designed to suit the differing needs of all women, would benefit everyone, since housing forms such a large part of the built environment which all see and experience, regardless of sex. Such a housing programme would not solve the problem of gender inequality in itself, but it would be an important building block to a better world.

Appendix

Table A.1   The composition of a group of 1,000 representative households (analysed by size of household and relationship pattern)

Household type	Total	Number of persons in household						
		1	2	3	4	5	6	7+
Single person	46[1]	46	-	-	-	-	-	-
Housewife + husband only	162[2]	-	162	-	-	-	-	-
Housewife + husband + sons or daughters aged 0–14 (There may be others)	372[3]	-	-	111	106	75	37	43
Housewife + husband + sons or daughters aged 15+ (no children aged 0–14. There may be other adults)	194[4]	-	-	79	66	32	10	7
All persons aged 15+ (but not single-person households consisting solely of housewife and husband, nor households containing sons or daughters of the housewife)	103[5]	-	29	47	18	5	3	1
Others	123[6]	-	28	26	23	20	15	11
All households	1,000	46	219	263	213	132	65	62

Source: Central Office of Information, Wartime Social Survey (1949) *The British Household*, London: Central Office of Information

Notes:
1. This is likely to be an underestimate. The correct figure is probably between 70 and 80.
2. In 31 of these households the housewife is under 35 years of age and in 94 she is aged 50 or over.
3.(a) About one in six of these households will contain grown-up sons or daughters or other adult relatives.
  (b) Only 13 of these households will contain boarders.
4. Only 8 of these households will contain boarders.
5.(a) In about one-third of these households the housewife has a husband.
  (b) 29 of these households will contain boarders, 9 consisting only of the housewife and a boarder, and 8 of the housewife, her husband and a boarder.
6.(a) About one-third of these households contain the housewife, her husband, married sons or daughters and grandchildren.
  (b) 10 of these households contain boarders.

Notes

The initials PRO and GLRO used in the notes below refer to the Public Records Office and the Greater London Records Office, respectively.

Chapter one   Introduction

1. See Breugel (1986) 'The Reserve Army of Labour, 1974–1979' in Feminist Review (ed.) (1986) *Waged Work: A Reader*, London: Virago, pp.40–53 for a full consideration of Beechey's argument.
2. Merrett & Gray (1982) *Owner Occupation in Britain*, London: Routledge & Kegan Paul, ch. 15, together with Saunders (1989) 'The Meaning of Home in Contemporary English Culture' *Housing Studies*, vol.4, no.3, pp.177–192, provide a full discussion.

Chapter two   Women as homemakers I

1. See illustration of cottage interior in Unwin & Parker (eds) (1901) *The Art of Building a Home*, London: Longmans, p.3.
2. See Pearson (1988) *The Architectural and Social History of Co-operative Living*, London: Macmillan, for a full account of experiments in co-operative living.

Chapter three   Women as homemakers II

1. *Picture Post* (1941) 'A Plan for Britain', *Picture Post* (special issue) vol.10, no.1, 4 January 1941.
2. Greater London Record Office (GLRO) Minutes of the LCC Housing Committee 24 January 1951 Item (31) Presented Papers vol.105, Jan–Feb 1951.
3. Public Record Office (PRO) HLG (101/252).
4. Lord Balfour, *Parliamentary Debates*, House of Lords, vol.140, 1945–1946, col. 766.
5. Viscount Samuel, *Parliamentary Debates*, House of Lords, vol.140, 1945–1946, col. 766.

6. GLRO Minutes of the LCC Housing Committee, 22 March 1950, Item (24) 'Points Scheme'. The Points Scheme was revised to take account of (a) bedroom deficiency, (b) bad conditions according to a finely graduated scale, (c) additional points for shared accommodation or shared or inadequate cooking facilities, water supply or sanitary arrangements and (d) families – each family member was awarded five points divided by the number of rooms occupied.

   In addition, between 10 and 20 would be awarded for certified ill health, separated families were given a further five points and families in rest centres or institutions were given five points.

   It would seem, given the bad housing conditions which then prevailed generally, that those most likely to benefit from this scheme would be families with children and chronically ill people.

7. GLRO Minutes of the LCC Housing Committee, 13 September 1944, Item (8), Presented Papers vol.77, 1944.

Chapter four   Women workers and the domestic ideal

1. GLRO CL/HSG/1/149 Miscellaneous, pp.4–5.
2. This view is contested, but only in passing, by Glynn & Oxborrow (1976) *Interwar Britain – Economic and Social History*, London: Allen & Unwin, p.110.
3. PRO CAB 117/144 Post-War Resettlement of Manpower.
4. PRO CAB 123/60 Proposals for Post-war Re-organisation of Industry 1942–1944.
5. Ibid.
6. PRO HLG 102/10 Distribution of Industry, 1945.
7. GLRO CL/HSG/1/61 Greater London Plan.
8. GLRO Housing Committee Presented Papers vol.102, May–June 1950, 21 June 1950, Item 11.
9. GLRO Housing Committee Presented Papers vol.102, May–June 1950, 7 June 1950, Item 30.

Chapter five   A woman's home is her factory

1. PRO HLG 37/63 Central Housing Advisory Committee Sub-Committee on Design of Houses and Flats: Circulated Papers.
2. Ibid.
3. Ibid.
4. Ibid.
5. For example, one London Borough, even in 1989, refuses to give planning permission to developments below the Parker Morris standard.
6. Interview by author with Baronness Denington, 3 January 1944
7. GLRO LCC Housing Committee Presented Papers vol.87, September 1947, Item 10.
8. Interview with Baronness Denington (see note 6).

9. GLRO LCC Housing Committee Presented Papers vol.126, 6 October 1954, Item 25.

10. GLRO Housing Committee, Housing Development and Management Sub-Committee, 4 May 1955, Item (14), Presented Papers vol.133, April–May 1955.

11. GLRO LCC Housing Committee, 20 April 1950, Item (17), Presented Papers vol.101, March–April 1950.

12. GLRO LCC File CL/HSG/1/67 Provision of Open Spaces.

13. GLRO LCC Housing Committee Presented Papers vol.87, February 1947, Item (13).

14. GLRO LCC Housing Committee Presented Papers vol.104, 22 November 1950.

15. GLRO LCC Housing Committee, 11 April 1951, Item 3 (31), Presented Papers vol.106, March–April 1951.

16. GLRO LCC Housing Committee, 23 May 1951, Item (20), Presented Papers vol.107, March–April 1951.

17. GLRO LCC Housing Committee, 23 May 1951, Item (20), Presented Papers vol.107, May–June 1951.

18. GLRO LCC Housing Committee, Housing and Development Management Sub-Committee, 26 November 1952, Item (14), Presented Papers vol.115, November–December 1952.

19. GLRO LCC Housing Committee Minutes, 3 December 1952, Item (14).

20. PRO CAB 87/37 Ministerial and Official Committee on Housing 1945: Papers.

21. PRO HLG 101/247 Revision of Housing Subsidies: Correspondence 1951–1952.

22. PRO HLG 101/252 Housing Finance, Costs, Rents and Subsidies 1942–1946.

23. Ibid.

24. GLRO LCC Housing Committee, 26 January 1954, Item (13).

25. GLRO LCC Housing Committee, 5 July 1950, Item (12), Presented Papers vol.102, July–September 1950.

26. GLRO CL/HSG/1/72 Building Costs, March 1950–December 1955.

27. GLRO LCC Minutes 1955, pp.521–522.

28. GLRO LCC Housing Committee, 5 July 1950, Item (12), Presented Papers vol.102, July–September 1950.

29. GLRO LCC Minutes, 12 July 1955, p.41.

Chapter six 'We saw it as a dream'

1. See, for example, the work of Lubetkin and Tecton: the Finsbury Health Centre provides a fine illustration of this point.

2. GLRO Presented Papers vol.80, 10 April 1946, Item 2(3).

3. See *Architects' Journal* vol.109, no.2822, 10 March 1949, pp.226–227; vol.109, no.2823, 17 March 1949, pp.251–254; vol.109, no.2832, 19 May 1949, pp.451–452.

4. Interview with Baroness Denington by author on 3 January 1984.

'The valuer used to order everything on a large site to be bulldozed so that nothing remained to prevent the maximum number of new housing units being put on the site.'

5. GLRO Minutes of the LCC, 20 December 1949, pp.779–781.
6. Interview with B. L. (Beak) Adams by author on 3 July 1984.
7. GLRO Minutes of the LCC Housing Committee, 24 January 1951, Item (31), Presented Papers vol.105, January–February 1951.
8. The first national council housing programme, instigated by the Addison Act of 1919, recommended the *Housing Manual 1919*, which was based on Unwin's ideas, as a guide.
9. *Architectural Review* vol.97, no.582, June 1945, see especially pp.166–170.
10. GLRO Minutes of the Joint Development Sub-Committee to the Housing Committee, 10 February 1954, Item (5), Presented Papers vol.124, January–February 1954.
11. Interview with B. L. (Beak) Adams (see note 6).
12. Ibid.

Chapter seven   A respectable life

1. Interview with B. L. (Beak) Adams by author on 3 July 1984.
2. For a fuller account of the interviews see Roberts (1986) unpublished thesis. The respondents' names and the names of the estate and surrounding roads have been changed in order to preserve the respondents' anonymity.
3. GLRO HSG/GEN/3/15 Rent Book and File of a 'Model Tenant'.
4. GLRO LCC Council Minutes, 16 November 1954, p.642.
5. Interview with B. L. (Beak) Adams, (see note 1).

Chapter Eight   From there to here

1. Central Statistical Office (1989) *Social Trends*, London: HMSO, p.144. Households with a gross income of up to £100 per week were spending, on average, £34.98 on mortgage payments, insurance and rates per week.

Bibliography

ABCA, Army Bureau of Current Affairs (1943a) 'Building the Post-War Home' by B. S. Townroe, *Current Affairs*, 20 November 1943, p.56.

ABCA, Army Bureau of Current Affairs (1943b) 'When the Lights Go On' by W. E. Williams, *Current Affairs*, 31 July 1943, p.48.

Abercrombie, P. & Forshaw, J. H. (1943) *The County of London Plan*, London: Macmillan.

Addison, P. (1977) *The Road to 1945*, London: Quartet.

Alexander, S. (1976) 'Women's Work in Nineteenth-Century London, A Study of the Years 1820–1850' in Mitchell, J. & Oakley, A. (eds) *The Rights and Wrongs of Women*, Harmondsworth: Pelican.

Allatt, P. (1983) 'Men and War: Status, Class and the Social Reproduction of Masculinity' in Gamarnikow, E. (ed.) *The Public and the Private*, London: Heinemann.

Allen, M. (1983) 'The Domestic Ideal and the Mobilization of Womanpower in World War II' *Women's Studies International Forum* 6(4): 401–412.

Anderson, M. (1979) 'The Relevance of Family History' in Harris, C. *The Sociology of the Family: New Directions for Britain*, Sociological Review Monograph 28, University of Keele.

Arnold, E. & Burr, L. (1985) 'Housework and the Application of Science' in Faulkner, W. & Arnold, E. (eds) *Smothered by Invention*, London: Pluto Press.

Ash, J. (1980) 'The Rise and Fall of High-rise Housing in England' in Ungerson, C. and Karn, V. *The Consumer Experience of Housing*, Farnborough: Gower.

Attfield, J. (1989) 'Inside Pram Town: A Case Study of Harlow House Interiors 1951–1961' in Attfield, J. & Kirkham, P. (eds) below.

Attfield, J. & Kirkham, P. (eds) (1989) *A View from the Interior: Feminism, Women and Design*, London: The Women's Press.

Ball, M. (1986) *Home Ownership: A Suitable Case for Reform*, London: Shelter.

Banham, R. (1960) *Theory and Design in the First Machine Age*, London: Architectural Press.

Banham, R. (1966) *The New Brutalism: Ethic or Aesthetic?* London: Architectural Press.

Barlow Report, The (1940) *The Report of the Royal Commission on the Distribution of the Industrial Population*, Cmnd 6153, London: HMSO.

Barrett, M. & McIntosh, M. (1980) 'The Family Wage: Some Problems for Socialists and Feminists' *Capital and Class* 11: 51–72.

Barrett, M. & McIntosh, M. (1982) *The Anti-Social Family*, London: Verso.

Barron, R. D. & Norris, G. M. (1976) 'Sexual Divisions and the Dual Labour Market' in Barker, D. L. & Allan, S. (eds) *Dependence and Exploitation in Work and Marriage*, London: Longmans.

Beattie, S. (1980) *A Revolution in London Housing: LCC Housing Architects and their Work 1893–1914*, London: Greater London Council & Architectural Press.

Beechey, V. (1986) 'Studies of Women's Employment' in Feminist Review (ed.) *Waged Work: A Reader*, London: Virago.

Bereano, P., Bose, C., & Arnold, A. (1985) 'Kitchen Technology and the Liberation of Women from Housework' in Faulkner, W. & Arnold, E. (eds) *Smothered by Invention*, London: Pluto Press.

Beveridge Report, The (1942) *Social Insurance and Allied Services*, Cmnd 6404, London: HMSO.

Beveridge, W. H. (1943) *'The Pillars of Security' and Other Wartime Essays and Addresses*, London: G. Allen & Unwin.

Black, M. (1985) *Food and Cooking in 19th Century Britain; History and Recipes*, Birmingham: Historic Buildings and Monuments Commission for England.

Bowley, M. (1945) *Housing And The State 1919–1944*, London: Allen & Unwin.

Boys, J. (1989) 'From Alcatraz to the OK Corral: Images of Class and Gender' in Attfield J. & Kirkham P. (eds) *A View from the Interior: Feminism, Women and Design*, London: The Women's Press.

Breugel, I. (1986) 'The Reserve Army of Labour 1974–1979' in Feminist Review (ed.) *Waged Work: A Reader*, London: Virago.

Brion, M. & Tinker, A. (1980) *Women in Housing: Access and Influence*, London: Housing Centre Trust.

Brown, G. W. & Harris, T. (1978) *Social Origins of Depression: A Study of Psychiatric Disorder in Women*, London: Tavistock.

Burnett, J. (1980) *A Social History of Housing 1815–1970*, London: Methuen.

Calder, A. (1965) *The People's War*, London: Jonathan Cape.

Central Housing Advisory Committee (1952) *Living in Flats*, London: HMSO.

Central Office of Information (1949) *The Social Survey: The British Household*, London: Central Office of Information.

Central Statistical Office (1989) *Social Trends*, London: HMSO.

CHAC Central Housing Advisory Committee, the Dudley Report (1944)
Design of Dwellings, London: HMSO.

Cleeve Barr, A. W. (1958) *Public Authority Housing*, London: Batsford.

Cockburn, C. (1983) *Brothers: Male Dominance and Technological Change*,
London: Pluto Press.

Comer, L. (1974) *Wedlocked Women*, Leeds: Feminist Books.

Cooper, I. (1978) 'Patterns of Space Usage in Primary Schools: An
Observational Study', PhD thesis, University of Wales.

Cullingworth, J. B. (1975) *Environmental Planning 1939–69, vol 1.
Reconstruction and Land Use Planning*, London: HMSO.

Cullingworth, J. B. (1979) *Essays on Housing Policy*, London: Allen and
Unwin.

Darley, G. (1975) *Villages of Vision*, London: Architectural Press.

Daunton, M. J. (1983) 'Public and Private Space: The Victorian City and the
Working-class Household' in Fraser, D. and Sutcliffe, A. (eds) *The Pursuit
of Urban History*, London: Edward Arnold.

Davidoff, L. (1976) 'The Rationalisation of Housework' in Barker, D.L. and
Allen, S. (eds) *Dependence and Exploitation in Work and Marriage*,
London: Longmans.

Davidoff, L., L'Esperance, J. & Newby, H. (1976) 'Landscape with Figures:
Home and Community in English Society' in Mitchell, J. and Oakley, A.
(eds) *The Rights and Wrongs of Women*, Harmondsworth: Penguin.

Davidoff, L. & Hall, C. (1987) *Family Fortunes: Men and Women of the
English Middle Class, 1780–1850*, London: Hutchinson.

Davies, A. (1984) *Where Did The Forties Go?*, London: Pluto Press.

Delphy, C. (1970) 'The Main Enemy: A Materialist Analysis of Women's
Oppression' in Delphy, C. (1984) *Close to Home*, London: Hutchinson.

Delphy, C. (1974) 'Sharing the Same Table: Consumption and the Family' in
Delphy, C. (1984), *Close To Home*, London: Hutchinson.

Denby, E. (1941) 'Plan the Home' *Picture Post*, 10, 1, 4, January 1941,
pp.21–24.

DoE (Department of the Environment) (1971) *Fair Deal for Housing*, Cmnd
4728, London: HMSO.

DoE (Department of the Environment) (1972) *Design Bulletin 25: The Estate
Outside The Dwelling*, London: HMSO.

DoE (Department of the Environment) (1973) *Design Bulletin 27: Children At
Play*, London: HMSO.

Douglas, M. (1966) *Purity and Danger: An Analysis of the Concepts of
Pollution and Taboo*, London: Routledge & Kegan Paul.

Drew, J. (1944) 'The Kitchen of the Future' *Women's Illustrated*,
5 February 1944.

Dudley Report, The (1944). See CHAC (1944).

Dunleavy, P. (1981) *The Politics of Mass Housing in Britain 1945–1975*,
London: Clarendon Press.

Durant, R. (1939) *A Watling: A Survey of Social Life on a New Housing Estate*, London: King.

Dyhouse, C. (1981) *Girls Growing Up in Late Victorian and Edwardian England*, London: Routledge & Kegan Paul.

Dyhouse, C. (1986) 'Mothers and Daughters in the Middle-class Home 1870–1914', in Lewis, J. (ed.) *Labour and Love: Women's Experience of Home and Family 1850–1914*, Oxford: Basil Blackwell.

Ehrenreich, B. & English, P. (1979) *For Her own Good*, London: Pluto Press.

Esher, L. (1981) *A Broken Wave: The Rebuilding of England 1940–1980*, London: Allen Lane.

Evans, R. (1978) 'Rookeries and Model Dwellings: English Housing Reform and the Moralities of Private Space' *Architectural Association Quarterly*, 10(3): 25–36.

Finer Report, The (1974) *Report of the Committee on One-parent Families, vols I & II*, London: HMSO.

Foot, M. (1973) *Aneurin Bevan*, London: Paladin.

Forty, A. (1975) 'The Electric Home' in *British Design*, Milton Keynes: Open University.

Forty, A. (1977) 'Housewives Aesthetic', *Architectural Review* 142, 969: 284–286 November 1977.

Frampton, K. (1985) *Modern Architecture: a Critical History*, London: Thames & Hudson.

Furneaux Jordan (1956) 'LCC New Standards in Official Architecture' *Architectural Review*, 120 (718): 302–324 November 1956.

Gallagher, P. (1982) 'Ideology and Housing Management' in English, J. (ed.) *The Future of Council Housing*, London: Croom Helm.

Gauldie, E. (1974) *Cruel Habitations*, London: Allen & Unwin.

George, V. & Wilding, P. (1978) *Ideology and Social Welfare*, London: Routledge & Kegan Paul.

Gibson, M. & Langstaff, M. (1982) *An Introduction to Urban Renewal*, London: Hutchinson.

Giedion, S. (1948) *Mechanisation Takes Command*, New York: Oxford University Press.

Ginsburg, N. (1979) *Class, Capital and Social Policy*, London: Macmillan.

Gittins, D. (1985) *The Family In Question: Changing Households and Familiar Ideologies*, Basingstoke: Macmillan Education.

Gittins, D. (1986) 'Marital Status, Work and Kinship 1850–1930' in Lewis, J. (ed.) *Labour and Love: Women's Experience of Home and Family 1850–1940*, Oxford: Basil Blackwell.

Gittus, E. (1976) *Flats, Families and the Under-fives*, London: Routledge & Kegan Paul.

Glynn, S. & Oxborrow, J. (1976) *Interwar Britain: a Social and Economic History*, London: Allen & Unwin.

Greater London Council (1985) *Changing Places*, London: Greater London Council.

Hall, C. (1982) 'The Butcher, the Baker, the Candlestickmaker: the shop and the family in the Industrial Revolution' in Whitelegg, E., Arnot, M., Bartels, E., Beechey, V., Birke, L., Himmelweit, S., Leonard, D., Ruel, S. & Speakerman, M. (eds) *The Changing Experience of Women*, Oxford: Martin Robertson.

Hall, P. G. (1962) *The Industries of London Since 1861*, London: Hutchinson University Library.

Halton, E. (1943) 'The Home of the Citizen' *British Way and Purpose*, 9, 1944, Directorate of Army Education.

Hartman, H. (1979) 'The Unhappy Marriage of Marxism and Feminism' in Sargent, L. (ed.) (1981) *The Unhappy Marriage of Marxism and Feminism*, London: Pluto Press.

Hayden, D. (1981) *The Grand Domestic Revolution*, London and Cambridge, Mass.: Massachusetts Institute of Technology Press.

Hinchcliffe, T. F. M. (1981) 'Highbury New Park' *London Journal* 7(1): 29–44.

H. M. Government (1944) *Employment Policy*, Cmnd 6527, London: HMSO.

H. M. Government (1947) *Economic Survey*, Cmnd 7046, London: HMSO.

Hooks, B. (1984) 'Feminism: A Movement to End Sexist Oppression' in Phillips, A. (ed.) (1987) *Feminism and Equality*, Oxford: Basil Blackwell.

Howard, E. (1898) *Garden Cities of Tomorrow*, Eastbourne: Attic Books (1985 edn).

Humphries, J. (1981) 'Protective Legislation, the Capitalist State and Working-class Men: the Case for the 1842 Mines Regulation Act' *Feminist Review* 7:1–33.

Hunt, P. (1980) *Gender and Class Consciousness*, London: Macmillan.

Jackson, A. (1973) *Semi-detached London*, London: Allen & Unwin.

Jeffreys, M. (1964) 'Londoners in Hertfordshire: the South Oxhey Estate' in Centre for Urban Studies, Report no. 3 (1964) *London, Aspects of Change*, London: Macgibbon & Kee.

Jencks, C. (1973) *Modern Movements in Architecture*, London: Architectural Press.

Land, H. (1976) 'Women: Supporters or Supported?' in Barker, D. L. and Allen, S. (eds) (1976) *Sexual Divisions and Society: Process and Change*, London: Tavistock.

Land, H. (1979) 'The Boundaries between the State and the Family' in Harris, C. (ed.) (1979) *The Sociology of the Family: New Directions for Britain*, Sociological Review Monograph 28: University of Keele.

Land, H. (1980) 'The Family Wage', *Feminist Review* 6:55–77.

Le Corbusier, (1929) *The City of Tomorrow*, London: Architectural Press.

Lewis, J. (1980) *The Politics of Motherhood*, London: Croom Helm.

Lewis, J. (1985) 'The Debate on Sex and Class' *New Left Review* 149:108–120.

Mackenzie, S. & Rose, D. (1983) 'Industrial Change, the Domestic Economy and Home Life' in Anderson, J., Duncan, S. & Hudson, R. (eds) *Redundant Spaces: Social Change and Industrial Decline in Cities and Regions*, London: Academic Press.

Mackintosh, H. J. M. (1952) *Housing and Family Life*, London: Cassel & Co.

MacFarlane, B. (1984) 'Homes Fit for Heroines: Housing in the Twenties' in Matrix (eds) *Making Space: Women and the Built Environment*, London: Pluto Press.

Martin, L. (1983) *Buildings and Ideas 1933–1983*, Cambridge: Cambridge University Press.

Marwick, A. (1970) *Britain In The Century Of Total War*, Harmonsworth: Penguin.

Marwick, A. (1982) *British Society Since 1945*, Harmondsworth: Pelican.

Marx, K. (1887) *Capital, Volume I*, London: Lawrence & Wishart.

Mass-Observation (1943) *An Enquiry into People's Homes*, London: Murray.

Merrett, S. (1979) *State Housing in Britain*, London: Routledge & Kegan Paul.

Merrett, S. & Gray, F. (1982) *Owner Occupation in Britain*, London: Routledge & Kegan Paul.

MoH (Ministry of Health) (1944) *Housing Manual 1944*, London: HMSO.

MoHLG (Ministry of Housing and Local Government) (1953) *Houses 1953*, London: HMSO.

MoHLG (Minstry of Housing and Local Government) (1961) *Homes For Today and Tomorrow*, Report of the Parker Morris Committee, London: HMSO.

Mumford, L. (1945) 'On the Future of London' *Architectural Review* 97 (577):3–10.

Murie, A. (1983) *Housing Inequality and Deprivation*, London: National Consumer Council.

Muthesius, S. (1982) *The English Terraced House*, London: Yale University Press.

Newby, H. (1980) *Green and Pleasant Land? Social Change in Rural England*, Harmondsworth: Penguin.

Oakley, A. (1976) *Housewife*, Harmondsworth: Penguin.

Oliver, P., Davis, I. & Bentley, I. (1981) *Dunroamin: The Suburban Semi and its Enemies*, London: Barrie & Jenkin.

Orbach, L. (1977) *Homes for Heroes*, London: Seely Service.

Pahl, R. E. (1984) *Divisions of Labour*, Oxford: Basil Blackwell.

Parker, B., & Unwin, R. (1901) *The Art of Building A Home*, London: Longmans Green.

Partridge, J. (1980) 'Alton West' in 'The Alton Estate, Roehampton After 25 Years' *Housing Review* 29 (5): 170–172 September–October 1980.

Pember Reeves, M. (1913) *Round About a Pound A Week*, reprinted 1979 London: Virago.

Pepper, S. (1981) 'Ossulton Street: Early LCC Experiments in High-rise Housing 1925–1929', *London Journal*, 7(1): 45–64.

Phillips, A. & Taylor, B. (1980) 'Sex and Skill' in Feminist Review (ed.) (1986) *Waged Work: A Reader*, London: Virago.

Power, A. (1987) *Property before People: The Management of Twentieth Century Council Housing*, London: Allen & Unwin.

Ravetz, A. (1974) *Model Estate*, London: Croom Helm.

Ravetz, A. (1980) *Remaking Cities*, London: Croom Helm.

Ravetz, A. (1984) 'The Home of Women: a View from the Interior' in Attfield, J. & Kirkham, P. (eds) (1989) *A View from the Interior: Feminism, Women and Design*, London: The Women's Press.

Reid, I., & Wormald, E. (eds) (1982) *Sex Differences in Britain*, London: Grant McIntyre.

Report of the Royal Commission on Population (1949), Cmnd 7695, London: HMSO.

Riley, D. (1983) *War in the Nursery: Theories of the Child and Mother*, London: Virago.

Roberts, M. (1984) 'Private Kitchens, Public Cooking' in Matrix (eds) *Making Space: Women and the Built Environment*, London: Pluto Press.

Roberts, M. (1986) 'The Modernisation of Family Life? Sexual Divisions in Architecture and Town Planning 1940–1947', unpublished PhD thesis: University of Wales.

Roberts, M. (1988) 'Caretaking – Who Cares?' in Markus, T., Teymur, N. & Woolley, T. (eds) *Rehumanising Housing*, London: Butterworth.

Saunders, P. (1989) 'The Meaning of Home in Contemporary English Culture' *Housing Studies* 4 (3): 177–192.

SCLRP (Standing Conference on London Regional Planning) (1945) *Greater London Plan*, London: HMSO.

Scott Report, The (1942) *Report of the Committee on Land Utilisation in Rural Areas*, Cmnd 6378, London: Tavistock.

Segal, L. (1987) *Is the Future Female?: Troubled Thoughts on Contemporary Feminism*, London: Virago.

Self, P. (1957) *Cities in Flood*, London: Faber & Faber.

Smithson, A. & Smithson, P. (1970) *Ordinariness and Light: Urban Theories 1952–1960*, London: Faber & Faber.

Spring Rice, M. (1939) *Working-Class Wives*, Harmondsworth: Pelican.

Stacey, M. (1970) *Tradition and Change: A Study of Banbury*, London: Oxford University Press (first published 1960).

Stedman Jones, G. (1971) *Outcast London: A Study in the Relationship between Classes in Victorian Society*, London: Oxford University Press.

Summerfield, P. (1984) *Women Workers in the Second World War*, London: Croom Helm.

Swenarton, M. (1981) *Homes Fit for Heroes*, London: Heinemann Educational.

Tarn, J. N. (1973) *Five Per Cent Philanthropy*, London: Cambridge University Press.

Taylor, B. (1983) *Eve and the New Jerusalem*, London: Virago.

Tilly, L. & Scott, J. (1978) *Women, Work and Family*, USA: Holt, Rinehart & Winston.

TCPA (Town and Country Planning Association) (1943) *Town and Country Planning* 11:120–121.

Tucker, J. (1966) *Honourable Estates*, London: Gollancz.

Uthwatt Report, The (1942) *Expert Committee on Compensation and Betterment: Final Report*, Cmnd 6386, London: HMSO.

Vanek, J. (1974) 'Time Spent in Housework' in Amsden, A. (ed.) (1980) *The Economics of Women and Work*, Harmondsworth: Penguin.

Walker, L. (1989) 'Women Architects' in Attfield, J. & Kirkham, P. (eds) *A View from the Interior: Feminism, Women and Design*, London: The Women's Press.

Walker, S. (1983) 'Women and Housing in Classical Greece: the Archaeological Evidence' in Cameron, A. & Kuhrt, A. (eds) *Images of Women in Antiquity*, London: Croom Helm.

Watson, S. (1986) *Housing and Homelessness: A Feminist Perspective*, London: Routledge & Kegan Paul.

White, J. (1981) *Rothschild Building: Life in an East End Tenement Block 1887–1920*, London: Routledge & Kegan Paul.

White, J. (1986) *The Worst Street in North London*, London: Routledge & Kegan Paul.

Williams, P. (1987) 'Constituting class and gender: a social history of the home, 1700–1901' in Thrift, N. and Williams, P. (eds) (1987) *Class and Space*, London: Routledge & Kegan Paul.

Williams, R. (1973) 'Base and Superstructure in Marxist Cultural Theory' in Williams, R. (1980) *Problems in Materialism and Culture*, London: Verso, pp.31–49.

Williams, R. (1975) *The Country and the City*, London: Paladin.

Wilmott, P. & Murie, A. (1988) *Polarisation and Social Housing*, London: Policy Studies Unit.

Wilson, E. (1977) *Women and the Welfare State*, London: Tavistock.

Wilson, E. (1980) *Only Halfway to Paradise: Women in Postwar Britain 1945–1968*, London: Tavistock.

Wohl, A. S. (1977) *The Eternal Slum*, London: Edward Arnold.

Women and Geography Study Group of the Institute of British Geographers (1984) *Geography and Gender*, London: Hutchinson.

Yorke, F. R. S. (1934) *The Modern Home*, London: Architectural Press.

Young, K. & Garside, P. (1982) *Metropolitan London: Politics and Urban Change 1837–1981*, London: Edward Arnold.

Young, M. & Wilmott, P. (1957) *Family and Kinship in East London*, Harmondsworth: Pelican.

Bibliography

Young, M. & Wilmott, P. (1973) *The Symmetrical Family*, London: Routledge & Kegan Paul.

Zaretsky, E. (1976) *Capitalism, the Family and Personal Life*, London: Pluto Press.

Zmroczek, C. (1984) 'Women's Work: Laundry in the Last Fifty Years', unpublished paper, Science Policy Research Unit: University of Sussex.

Name index

Subject index